CRASH COURSE on US PATENT LAW

by Arti Kane

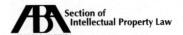
Section of
Intellectual Property Law
AMERICAN BAR ASSOCIATION

Cover Design: Amy Mandel, ABA Section of Intellectual Property Law;
Best Design Chicago, Inc.

Printed in the United States of America.

20 19 18 17 16 5 4 3 2 1

Library of Congress Cataloging-in-Publication Data

Names: Kane, Arti, author. | American Bar Association. Section of
    Intellectual Property Law, sponsoring body.
Title: Crash course on U.S. patent law / Arti Kane.
Other titles: Crash course on United States patent law
Description: Chicago : American Bar Association, 2016.
Identifiers: LCCN 2016003933 | ISBN 9781634253949
Subjects: LCSH: Patent laws and legislation—United States.
Classification: LCC KF3114.85 .K36 2016 | DDC 346.7304/86—dc23
LC record available at http://lccn.loc.gov/2016003933

Discounts are available for books ordered in bulk. Special consideration is given to state bars, CLE programs, and other bar-related organizations. Inquire at Book Publishing, ABA Publishing, American Bar Association, 321 North Clark Street, Chicago, Illinois 60654-7598.

www.ababooks.org

Table of Contents

A Crash Course on U.S. Patent Law

Are you an inventor, law student, patent engineer, patent agent, or patent attorney? Great, keep reading; this book was designed for you. This book is simply written in layman's terms so that not only the expert patent attorney but also the novice inventor can understand. I know that there are numerous books on patent law in the market; this book is by no means a substitute for them and should be read in conjunction with them. Why read this book? It is the only book on the market that provides concrete examples, including boilerplate language generated from my patent experience that is not found on the Internet. I wish I'd had a book like this when I first started my patent career. We will start with an introduction on patent law. Next, we will talk about the requirements for patentability: novelty and non-obviousness. Next, it will be useful to focus on 101 patentability. Then we will discuss claim drafting and prior art and prosecution history search, followed by the contents of a provisional application and patent application or specification. We will then talk about responses to office actions or amendments. Finally, it might be useful to dive into opinions, followed by the conclusion.

About the Author

In 1994, Arti Kane earned a degree in electrical engineering from the University of Sherbrooke in Sherbrooke, Quebec, Canada. In 1998, she graduated with degrees in civil and common law (cum laude) from the University of Ottawa in Ottawa, Ontario, Canada. In 2002, she became a member of the New York Bar and passed the U.S. Patent Bar. She has worked as a patent attorney for 15 years at prestigious law firms in Los Angeles, New York, San Diego, and Seoul, South Korea. She was also a member of the Ottawa Law Review (1996–1997) and Patent Committee of New York City Bar (2005–2006). She is presently a member of the Intellectual Property Institute of Canada and the Editorial Board of *Canadian Intellectual Property Review*. She also holds a copyright on a telecommunications project ("C" programming). She is fluent in French and has won many writing contests in French and English.

Introduction on Patent Law

A patent is a grant of an intellectual property right. Patents encourage new inventions and improvements to previous inventions. A patent holder has exclusive rights over a patent.

TYPES

- The United States Patent and Trademark Office grants three types of patents for inventions: utility, design, and plant patents.

RIGHTS

- A patent holder has the right to exclude others from making, using, or selling an invention. Exclusive rights to utility and plant patents endure for 20 years from the patent application date, while rights to a design patent endure for 14 years from the date of the patent grant.

INFRINGEMENT

- Infringement on a patent occurs when another person makes, uses, sells, or offers to sell a protected invention without the consent of the patent holder.

SEARCH

- Before applying for a patent, an applicant should conduct a search of previous patents to determine whether or not the invention has previously been patented.

FEES

- To maintain a patent, the patent holder must pay maintenance fees every 3 1/2, 7 1/2, and 11 1/2 years from the date of the patent grant. The fee amount varies depending on the type of patent and the fee schedule.

Patent Novelty

35 U.S.C. § 102—CONDITIONS FOR PATENTABILITY; NOVELTY

(a) Novelty; Prior Art.—A person shall be entitled to a patent unless—

(1) the claimed invention was patented, described in a printed publication, or in public use, on sale, or otherwise available to the public before the effective filing date of the claimed invention; or

(2) the claimed invention was described in a patent issued under Section 151, or in an application for patent published or deemed published under Section 122(b), in which the patent or application, as the case may be, names another inventor and was effectively filed before the effective filing date of the claimed invention.

(b) Exceptions.—

(1) Disclosures made 1 year or less before the effective filing date of the claimed invention.—A disclosure made 1 year or less before the effective filing date of a claimed invention shall not be prior art to the claimed invention under subsection (a)(1) if—

(A) the disclosure was made by the inventor or joint inventor or by another who obtained the subject matter disclosed directly or indirectly from the inventor or a joint inventor; or

(B) the subject matter disclosed had, before such disclosure, been publicly disclosed by the inventor or a joint inventor or another who obtained the subject matter disclosed directly or indirectly from the inventor or a joint inventor.

(2) Disclosures appearing in applications and patents.— A disclosure shall not be prior art to a claimed invention under subsection (a)(2) if—

(A) the subject matter disclosed was obtained directly or indirectly from the inventor or a joint inventor;

(B) the subject matter disclosed had, before such subject matter was effectively filed under subsection (a)(2), been publicly disclosed by the inventor or a joint inventor or another who obtained the subject matter disclosed directly or indirectly from the inventor or a joint inventor; or

(C) the subject matter disclosed and the claimed invention, not later than the effective filing date of the claimed invention, were owned by the same person or subject to an obligation of assignment to the same person.

(c) Common Ownership Under Joint Research Agreements.— Subject matter disclosed and a claimed invention shall be deemed to have been owned by the same person or subject to an obligation of assignment to the same person in applying the provisions of subsection (b)(2)(C) if—

(1) the subject matter disclosed was developed and the claimed invention was made by, or on behalf of, 1 or more parties to a joint research agreement that was in effect on or before the effective filing date of the claimed invention;

(2) the claimed invention was made as a result of activities undertaken within the scope of the joint research agreement; and

(3) the application for patent for the claimed invention discloses or is amended to disclose the names of the parties to the joint research agreement.

(d) Patents and Published Applications Effective as Prior Art.— For purposes of determining whether a patent or application for patent is prior art to a claimed invention under subsection (a)(2), such patent or application shall be considered to have been effectively filed, with respect to any subject matter described in the patent or application—

> **(1)** if paragraph (2) does not apply, as of the actual filing date of the patent or the application for patent; or

> **(2)** if the patent or application for patent is entitled to claim a right of priority under Section 119, 365(a), or 365(b), or to claim the benefit of an earlier filing date under Section 120, 121, or 365(c), based upon 1 or more prior filed applications for patent, as of the filing date of the earliest such application that describes the subject matter.

Patent law in the United States bestows legal rights upon a person who invents a new or useful process or device. Parties who challenge a patent may do so by denying the device is, in fact, new. They support their arguments with writings, blueprints, or other documents, with known dates that can establish that the process or device was already publicly known before the filing.

THE ALL-ELEMENTS RULE

Courts in the United States interpret the novelty requirement so that a claim will fail if a single document can be produced that has already described the invention within its "four corners." This means that somebody before the patent applicant previously put all the critical pieces together. For example, an applicant for a patent proposes an invention with elements A, B, and C. The challenging party asserts that three years previously Jane Doe discussed elements A and B of this invention, and at about the same time, in a separate writing, author Joe Smith discussed element C. This argument does not undermine the novelty requirement because, in this example, neither Doe nor Smith has all three elements together within the same "four corners."

PRIOR ART AND SIMILARITY

The body of publicly available discussion of an invention and possible modifications prior to the filing of a patent application are known as the "prior art" of that device or process. The parties who are challenging a patent find that there was something similar available in the public domain. This leads to some closely contested disputes.

GRACE PERIOD

But what if the inventor has created the "prior art" herself? For example, Jane Doe has been working on an improved product for many years. Thirteen months ago, she negligently allowed others to copy her preliminary designs. Now, she has applied for a patent. Her application could be rejected because it is not novel, on the basis of the prior art created by the applicant herself. The inventor has already put her invention, as embodied in that prior art, into the public domain. But if the disclosure occurred less than one year before the application was filed, it is within the statutory grace period.

Patent Non-Obviousness

35 U.S.C. § 103—CONDITIONS FOR PATENTABILITY; NON-OBVIOUS SUBJECT MATTER

A patent for a claimed invention may not be obtained, notwithstanding that the claimed invention is not identically disclosed as set forth in Section 102, if the differences between the claimed invention and the prior art are such that the claimed invention as a whole would have been obvious before the effective filing date of the claimed invention to a person having ordinary skill in the art to which the claimed invention pertains. Patentability shall not be negated by the manner in which the invention was made.

Non-obviousness rejections are very common in prosecution of patent applications before the United States Patent and Trademark Office. Thus, this somewhat subjective requirement is a key issue in almost every patent case. An examiner is not allowed to simply conclude that an invention is obvious; instead, an examiner must show specific documents that set out the various elements of the invention, along with some reason to combine the different documents to arrive at the invention. On top of that, examiners are given a very limited time to do a search for the documents. Moreover, it is the burden of the examiner to show an invention is obvious. For that reason inventions that may seem obvious to a layperson are sometimes issued patents.

TEACHING–SUGGESTION–MOTIVATION TEST

Further, the combination of previously known elements can be considered obvious. As stated by *Winner Int'l Royalty Corp. v. Wang*, 202 F.3d

1340, 1348 (Fed. Cir. 2000), there must be a suggestion or teaching in the prior art to combine elements shown in the prior art in order to find a patent obvious. Thus, in general, the critical inquiry is whether there is something in the prior art to suggest the desirability, and thus the obvious nature, of the combination of previously known elements.

This requirement is generally referred to as the "teaching–suggestion–motivation" (TSM) test and serves to prevent hindsight bias. *In re Kahn*, 441 F.3d 977, 989 (Fed. Cir. 2006). As almost all inventions are some combination of known elements, the TSM test requires a patent examiner (or accused infringer) to show that some suggestion or motivation exists to combine known elements to form a claimed invention. Some critics of the TSM test have claimed that the test requires evidence of an explicit teaching or suggestion to make a particular modification to the prior art, but the Federal Circuit has made clear that the motivation may be implicit, and may be provided, for example, by an advantage resulting from the modification. In other words, an explicit prior art teaching or suggestion to make a particular modification is sufficient, but not required for a finding of obviousness.

The TSM test has been the subject of much criticism. The U.S. Supreme Court addressed the issue in *KSR Intl. Co. v. Teleflex Inc.*, 127 U.S. 1727 (2007). The unanimous decision, rendered on April 30, 2007, overturned a decision of the Federal Circuit and held that it "analyzed the issue in a narrow, rigid manner inconsistent with Section 103 and our precedents," referring to the Federal Circuit's application of the TSM test. The court held that, while the ideas behind the TSM test and the Graham analysis were not necessarily inconsistent, the true test of non-obviousness is the Graham analysis. However, according to former Chief Judges Michel and Rader, the TSM test remains a part of the Federal Circuit's analysis, though it is applied mindful of the decision in *KSR*. A KSR-style obviousness analysis was used in *Perfect Web Technologies, Inc. v. InfoUSA, Inc.*, 587 F.3d 1324 (Fed. Cir. Dec. 2, 2009), to invalidate a patent due to the obvious nature of the asserted claims.

Graham Factors

The factors a court will look at when determining obviousness in the United States were outlined by the Supreme Court in *Graham et al. v. John Deere Co. of Kansas City et al.*, 383 U.S. 1 (1966) and are commonly referred to as the "Graham factors." The court held that obviousness should be determined by looking at

1. the scope and content of the prior art,
2. the level of ordinary skill in the art,
3. the differences between the claimed invention and the prior art, and
4. objective evidence of non-obviousness.

In addition, the court outlined examples of factors that show "objective evidence of non-obviousness." They are

1. commercial success,
2. long felt but unsolved needs, and
3. the failure of others.

101 Patentability

35 U.S.C. § 101—INVENTIONS PATENTABLE

Whoever invents or discovers any new and useful process, machine, manufacture, or composition of matter, or any new and useful improvement thereof, may obtain a patent therefor, subject to the conditions and requirements of this title.

Section 101 patentability has been repeatedly asserted to disqualify and thus invalidate patents that are not specifically limited to some tangible medium, like a computer, or that fall within the categories of an abstract idea, natural phenomenon, or natural law. Most patents facing a 101 challenge are software and business-method patents. Courts have been increasingly inconsistent in defining what renders patent claims too "abstract" to survive section 101 scrutiny.

Functional claiming refers to claiming an invention by reciting what it does (its functions) rather than reciting its structure. Nothing in the patent statute prohibits an invention from being claimed using "functional" language. But a review of the case law suggests that the use of the so-called functional language in a patent claim may increase the likelihood that the claim will be held unpatentable or invalid.

Claim Drafting

The claims of a patent specification define the scope of protection of a patent, granted by the patent. The claims describe the invention in a specific legal style, setting out the essential features of the invention in a manner to clearly define what will infringe the patent. Claims are often amended during prosecution to narrow or expand their scope.

The claims may contain one or more hierarchical sets of claims, each having one or more main, independent claim setting out the broadest protection, and a number of dependent claims, which narrow that protection by defining more specific features of the invention.

ANATOMY OF AN INDEPENDENT CLAIM

There are many styles of *patent claim*. Here is a generic example of a modern-style *independent claim* commonly used in the mechanical and electrical arts.

1. A [class], comprising:

   a [element] for [function],

   a [element] [connection] the [element above] for [function],

   a [element] [connection] the [element above] for [function], and

   a [element] [connection] the [element above] for [function].

Now for an example. Suppose I've invented a new kind of remote control for a high-definition DVD player. Here's how I would draft an independent claim.

   class = remote control for media playback

I want broad coverage, and I don't think any kind of remote control (TV, CD player, etc.) has seen this improvement before. On the other hand, I don't want the examiner to dig up model airplane remote controls and use those against me, which might happen if I simply said "remote control."

element = housing

This is a good one to start with because it gives me a place to put everything else. Claim drafting is made easier when you begin with a housing, frame, semiconductor substrate, or other large, essential element.

element = dial

Dials exist in the prior art, but has anyone ever put one on a remote control before? I hope not, because this is my invention. I may have more than one dial, but the minimum requirement would be one, so this is why I say "a." Had I needed two, I would've said "a plurality of dials."

element = circuit

element = electromagnetic radiation source

Note the generic language. This could be anything from a light bulb to a UV-emitting LED.

Here is what my claim looks like:

1. A remote control for media playback, comprising:

a housing for being held in a user's hand,

a dial rotatable with respect to the housing for being turned by the user,

a circuit electrically coupled to the dial for generating a signal based on a position of the dial, and

an electromagnetic radiation source electrically coupled to the circuit for emitting electromagnetic radiation based on the signal generated by the circuit.

Using this claim as an example, draft a claim on the peanut butter and jelly sandwich.

Answer:

A sandwich comprising:

a first piece of bread having a first side and a second side with a dab of peanut butter spread on the first side of the first piece, and a second piece of bread having a first side and a second side with a bit of jelly spread on the first side of the second piece,

wherein the first and second pieces of bread are placed in proximity to each other such that the peanut butter touches the jelly.

OR

A method of making a sandwich comprising:

spreading a dab of peanut butter onto a first piece of bread,

spreading a bit of jelly onto a second piece of bread,

putting together the first and the second pieces of bread so that the peanut butter and the jelly are touching.

Prior Art and Prosecution History Search

Prior art (also known as *state of the art*, which also has other meanings, or *background art*) in most systems of patent law, constitutes all information that has been made available to the public in any form before a given date that might be relevant to a patent's claims of originality. If an invention has been described in the prior art, a patent on that invention is not valid.

HOW TO DO A PATENT SEARCH ON THE USPTO WEBSITE

You can search the USPTO's patent database to see if a patent has already been filed or granted that is similar to your patent.

1. Go to www.uspto.gov.
2. Click on Patents.
3. Click on Patent Search (Search for Patents).
4. Click on USPTO Patent Full-Text and Image Database (PatFT).
5. Under Searching Full Text Patents (Since 1976) click on Quick Search.
6. Enter keywords that describe your invention.

Find prior art issued before the date of your invention.

Find prior art for U.S. Patent 8,414,072 (Collapsible Hanging Chair)

Term 1: collapsible chair

Searching U.S. Patent Collection . . .

Results of Search in U.S. Patent Collection db for:

"**collapsible chair**": 320 patents.
Hits 1 through 50 out of 320

Next 50 Hits

Jump To

Refine Search

	PAT. NO.		Title
1	8,967,578	**T**	Method and apparatus for swivel device
2	8,905,471	**T**	Collapsible chair with table
3	8,899,686	**T**	Collapsible chair for leisure
4	8,894,139	**T**	Collapsible lightweight hammock chair
5	8,886,287	**T**	Tissue-stabilization device and method for medical procedures
6	8,876,203	**T**	Collapsible chair
7	8,870,294	**T**	Collapsible chair
8	8,864,222	**T**	Unifoldable reclining chair
9	8,857,825	**T**	Ice fish house base and expansion lever

10	8,851,503	T	Dual-chair beach wagon
11	8,833,000	T	Continuous tension, discontinuous compression systems and methods
12	8,814,261	T	Adirondack chair with double fulcrum
13	8,801,095	T	Deployable chair for stroller coupling
14	8,794,699	T	Collapsible chair having foldable arm tray
15	8,794,698	T	Collapsible video gaming chair
16	8,764,105	T	Offset pyramid hinge folding chair
17	D707,914	T	Creeper with a collapsible chair
18	8,733,786	T	Recreational cart assembly
19	8,727,655	T	Folding chair safety lock
20	8,721,005	T	Collapsible chair
21	8,690,470	T	Adjustable connector for tubular frames
22	8,690,236	T	Reconfigurable collapsible chair
23	8,616,630	T	Tailgating picnic bench and table combination
24	8,544,809	T	Method and apparatus for swivel device
25	8,511,747	T	Collapsible chair with collapsible back support
26	8,479,663	T	Portable entertainment system
27	8,449,026	T	Convertible seating assembly
28	8,388,056	T	Heated collapsible article of furniture
29	8,381,749	T	Adjustable hunting blind
30	8,344,272	T	Chair-based scale with a movable load-cell housing for transitioning between a weighing and a nonweighing configuration
31	8,328,284	T	Lock release assembly for a collapsible chair having a fold-down back
32	8,322,784	T	Collapsible chair
33	8,303,032	T	Portable collapsible chair and sling
34	8,292,361	T	Collapsible chair
35	8,262,157	T	Hinge collapsible portable slat seat

36	8,256,042	T	Collapsible bed frame including cross units and method for constructing collapsible bed frame
37	8,251,452	T	Gaming chairs
38	8,251,442	T	Two-way foldable chair
39	8,186,755	T	Collapsible canopy along with article of furniture and method incorporating the same
40	8,162,349	T	Collapsible carrier
41	8,141,944	T	Collapsible chair having reduced linkages
42	8,100,469	T	Collapsible chair with curved back support
43	8,087,688	T	Collapsible pushchair
44	8,056,982	T	Compactable, collapsible chair
45	7,963,531	T	Collapsible utility cart
46	D639,124	T	Fabric cup holder for collapsible chair
47	7,950,744	T	Collapsible integral foot rest
48	7,950,673	T	Stair chair
49	7,942,476	T	Collapsible chair
50	7,883,143	T	Collapsible portable child seat

HOW TO DO A PATENT SEARCH IN GOOGLE PATENTS

Google Patents is a search engine from Google that indexes patents and patent applications from the United States Patent and Trademark Office (USPTO), European Patent Office (EPO), and World Intellectual Property Organization (WIPO).

1. Go to www.google.com/patents.
2. Type in the patent number or search term.
3. For example:

8,414,072

Collapsible hanging chair

HOW TO DO A PROSECUTION HISTORY SEARCH ON THE USPTO WEBSITE USING PUBLIC PAIR

The Patent Application Information Retrieval (PAIR) system provides IP customers a safe, simple, and secure way to retrieve and download information regarding patent application status. Public PAIR provides access to issued patents and published applications.

1. Go to www.uspto.gov.
2. Click on Patents.
3. Under Application Process, click on Checking application status.
4. Under Public PAIR, click on Check the status of a published application.
5. Enter the RECAPTCHA text.

   For example, under Search for Application, enter patent number 8,414,072.

6. Click on Image File Wrapper to see the prosecution history for the collapsible hanging chair.

Provisional Application

Before we talk about the patent specification, let us have a discussion on the provisional application and how it relates to the ordinary patent application. A provisional application is a document filed in the United States Patent and Trademark Office (USPTO) that establishes an early filing date, but does not mature into an issued patent unless the applicant files a regular non-provisional patent application within one year.

Although there is no required format for a provisional application, a provisional application typically includes a specification (description) and drawing(s) of an invention (drawings are required where necessary for the understanding of the subject matter sought to be patented), but does not require a summary or an abstract, claims, inventors' oaths or declarations, or any Information Disclosure Statement (IDS).

Patent Application or Specification

What is a patent application or specification?

A patent specification is a document that describes the invention for which a patent is sought and sets out the scope of the protection of the patent. As such, a specification generally contains a section that details the background and overview of the invention, a description of the invention and embodiments of the invention and claims, which set out the scope of the protection. A specification may include figures to aid the description of the invention, gene sequences and references to biological deposits, or computer code, depending upon the subject matter of the application. Most patent offices also require that the application includes an abstract, which provides a summary of the invention to aid searching. A title must also generally be provided for the application.

USING U.S. PATENT 8,414,072 (COLLAPSIBLE HANGING CHAIR) AS AN EXAMPLE, WRITE A PATENT SPECIFICATION ON THE PEANUT BUTTER AND JELLY SANDWICH.

In order to understand the anatomy of a patent application, we will analyze a few patents by emphasizing their boilerplate language. I have chosen the following patents because they are easy to understand and are all related to each other.

United States Patent **[PATENT NUMBER]**

[INVENTOR] **[ISSUE DATE]**

[INSERT TITLE]

[THE BOILERPLATE LANGUAGE IS INDICATED IN BOLD. YOU CAN USE THIS AS A TEMPLATE WHEN DRAFTING OTHER PATENT APPLICATIONS.]

Description

FIELD OF THE INVENTION

[DESCRIBE THE GENERAL FIELD IN WHICH THE PRESENT INVENTION IS LOCATED.]

The invention relates to equalization of high-speed digital communication channels using a threshold multiplexing feedback digital filter.

BACKGROUND

[DESCRIBE THE PRIOR ART AND EXPLAIN HOW THE PRESENT INVENTION SOLVES THE PROBLEMS OF THE PRIOR ART.]

In high-speed digital communication systems, communication channels often suffer from intersymbol interference (ISI). In such systems, coherent detection and equalization are necessary to achieve satisfactory performance. Equalizations are typically done using either linear digital filters such as finite impulse response (FIR) filters or non-linear digital filters such as decision feedback equalization (DFE) filters. Equalization can also be done using analog filters before sampling occurs.

Theoretically, FIR filters can be used to approximate any time-invariant impulse response with a large number of taps. The DFEs, which remove both pre- and post-cursor ISI using a feed-forward FIR filter followed by a feedback infinite impulse response (IIR) filter and a decision non-linearity, are reported to have better equalization results. However, most digital equalizers reported so far are very expensive to implement in very large scale integration (VLSI) systems due to the requirements of large device and silicon area count for high throughput data processing and high-speed analog to digital (A/D) conversion. **The analog equalizer solutions, on the other hand**, can be used to significantly reduce the equalizer device count. However, they suffer from poor design flexibility, reusability, testability, and manufacturability properties of the analog VLSI circuits.

SUMMARY

[USE SIMILAR WORDING TO THE ABSTRACT AND ADD "IN ACCORDANCE WITH THE INVENTION, THERE IS DISCLOSED."]

In accordance with the invention, there is disclosed an apparatus including a plurality of quantizers **each configured to compare** a selected threshold signal with an input signal and generate an output, a multiplexer, **coupled to** the plurality of quantizers, **that selects** one of the plurality of quantizer outputs according to a frequency response, and a multiplication-accumulation (MAC) unit, coupled to the multiplexer, the MAC to generate an output based on a previously selected one of the quantizer outputs according to the frequency response.

BRIEF DESCRIPTION OF THE DRAWINGS

The accompanying drawings are included to provide a further understanding of the invention, and are incorporated in and constitute a part of this specification. The drawings illustrate embodiments of the invention and, together with the description, serve to explain the principles of the invention. In the drawings,

[DESCRIBE EACH TYPE OF FIGURE.]

FIG. 1 is graphical representation or bode plot of frequency response H(w) plotted versus frequency (w);

FIG. 2 is a schematic diagram of a threshold multiplexing feedback digital equalizer architecture **in accordance with an embodiment of the invention**;

FIG. 3 illustrates a 1.sup.st order IIR filter frequency response **in accordance with an embodiment of the invention**;

FIG. 4 illustrates a 2.sup.nd order IIR filter frequency response **in accordance with an embodiment of the invention**;

FIG. 5 illustrates a 3.sup.rd order IIR filter frequency response **in accordance with an embodiment of the invention**;

FIG. 6 illustrates a 4.sup.th order IIR filter frequency response **in accordance with an embodiment of the invention**;

FIG. 7 is a general block diagram of a communication system wherein a second communication terminal comprises a filter which converts the analog voiceband signal into digital baseband data, **in accordance with an embodiment of the invention**.

DETAILED DESCRIPTION

[GIVE AN EXPLANATION OF THE FIGURES IN SEQUENCE AS IF YOU ARE TELLING A STORY. REMEMBER THE BOILERPLATE LANGUAGE IS INDICATED IN BOLD.]

[DESCRIBE THE ADVANTAGES OF THE PRESENT INVENTION IN THE FIRST PARAGRAPH OF THE DETAILED DESCRIPTION.]

A method and apparatus pertaining to equalization of high-speed digital communication channels using threshold multiplexing feedback digital filter **is described. In one embodiment**, a digital equalizer based on a feedback threshold multiplexing IIR filter structure is presented. An A/D converter is not required for this equalizer because a parallel quantization mechanism is included in the digital filter itself. In addition, only two-level data from the delayed output sequence are needed for the feedback data processing. **Consequently, this equalizer can be implemented in a smaller area and operated at higher speed than prior art configurations, and is very suitable for low cost VLSI circuit realization.**

In one aspect, the invention is concerned with retrieving communication signals over a communication line, such as a communication cable. **FIG. 1 shows a graphical representation** or bode plot of frequency response H(w) plotted versus frequency (w). In order to resemble an ideal cable where the impedance is matched and the slew rate is adjusted, it is necessary to increase the bandwidth of the cable. This can be achieved by using a high-pass filter, which will vary attenuation 5 of frequency response 2, **as shown in FIG. 1**. As a result, the bandwidth will increase from point A to point B, to form frequency response 3, **as shown in FIG. 1**.

In this invention, a digital equalizer based on a feedback threshold multiplexing IIR filter structure is presented. The general form of the difference equation for the IIR system is given by:

$$y.sub.k = x.sub.k - v.sub.th\ (c.sub.1\ y.sub.k - 1 + c.sub.2\ y.sub.k - 2 + \ldots + c.sub.n\ y.sub.k - n).ident.x.sub.k - a.sub.1\ y.sub.k - 1 - a.sub.2\ y.sub.k - 2 - \ldots - a.sub.n\ y.sub.k - n\ (1)$$

This equation relates the present output value with present values of the input and the present and past values of the output. Thus, the filtering process involves a recursive process using present and past values of the input. In equation (1), "y" represents the output signal, "x" represents the input signal, "c" represents a coefficient selected according to the frequency response desired, and "a.sub.n" represents an order coefficient representative of the frequency response where "n" is the order of the frequency response.

FIG. 2 schematically illustrates an embodiment of a threshold multiplexing feedback digital equalizer **according to the invention. As shown in FIG. 2**, equalizer 10 consists of parallel quantizer 20, multiplexer 25, digital delay line 35, and programmable multiplication-accumulation (MAC) unit 40. The ISI distorted signal {x.sub.k}, introduced into equalizer 10, is directly quantified by "m" parallel single-bit quantizers 20 with dedicated quantization thresholds {v.sub.i, = 1, 2, ... m} 15. Parallel quantizers 20 are, for example, comparators that function as 1-bit A/D converters because they recognize a two-level sequence, either high or low. The "m" two-level sequences are filtered using "m-to-1" multiplexer 25 to generate the two-level output sequence (either high or low).

The input signal {x.sub.k} is an analog signal and a different threshold signal {v.sub.i} 15 is associated with each quantizer 20 and comprises a value selected to be within the range of the input signal. For example, if the input signal {x.sub.k} is 1V, then the equally spaced threshold signals will range from $-1V$ to $+1V$ or $-1/2V$ to $+1/2V$, etc.

Each of the plurality of quantizers 20 comprises a comparator that subtracts threshold signal {v.sub.i} 15 from input signal {x.sub.k} to generate a representative digital output corresponding to whether threshold signal {v.sub.i} 15 is greater than or less than input signal {x.sub.k}. **In this embodiment**, if threshold signal {v.sub.i} 15 is less than input signal {x.sub.k}, then the representative digital output is high.

The output sequence which represents the complete history (past values) of the digital output signal, is obtained using flip-flops 30 on digital delay line 35 to delay each output signal by one clock cycle. The delayed two-level output data are used to create the control sequence of multiplexer 25 through MAC 40 and the outputs {f.sub.k} of MAC 40 are decoded such that at each sampling point there is a proportional mapping between "f.sub.k" and the threshold "v.sub.i" of the selected quantified output.

FIGS. 3, 4, 5, and 6 show several normalized filter frequency responses of filter expressed by equation (1) for some simple MAC 40 coefficient sets. As can be seen, a wide range of frequency responses can be constructed even with a limited number of feedback taps. **FIG. 3 illustrates** three different 1.sup.st order frequency responses, according to three different values for "a.sub.1." **FIG. 4 illustrates** three different 2.sup.nd order frequency responses, **according to** three sets of values for "a.sub.1" and "a.sub.2." **FIG. 5 illustrates** three different 3.sup.rd order frequency responses, according to three sets of values for "a.sub.1," "a.sub.2," and "a.sub.3." **Finally, FIG. 6 illustrates** three different 4.sup.th order frequency responses, **according to** three sets of values for "a.sub.1," "a.sub.2," "a.sub.3," and "a.sub.4."

MAC 40 multiplies the individual outputs by their respective coefficients c.sub.n and sums these outputs. The c coefficients are selected according to the frequency response desired since (v.sub.th • c.sub.n = a.sub.n) where "n" represents the order of the frequency response, v.sub.th the selected threshold signal, and a.sub.n an order coefficient representative of the frequency responses.

Multiplexer 25 selects one of the plurality of quantizer outputs according to the output generated from MAC unit 40. Mathematically, this proportional mapping is equivalent to generating an output sequence by subtracting a value which is proportional to MAC 40 output from the input sequence:

Varying the quantization thresholds {V.sub.i} 15 or c.sub.n, will provide desired coefficients "a.sub.n," which will, in turn, provide the frequency response desired, as explained earlier. This filter is non-linear because of in-loop quantization. An IIR digital filter can approximate the characteristics of equalizer 10 with the z-domain transfer function as:

##EQU1##

By performing the following steps on equation (1), equation (2) can be obtained:

##EQU2##

[THIS PARAGRAPH CAN BE USED FOR OTHER COMMUNICATION SYSTEMS.]

FIG. 7 is a general block diagram that illustrates a communication system 45 suitable for an implementation of the equalizer of the invention. System 45 can be implemented as part of a communication link between a transmitting signal and a receiving signal, such as in a Community Access Television (CATV) network, the Public Switched Telephone Network (PSTN), the Integrated Services Digital Network (ISDN), the Internet, a local area network (LAN), a wide area network (WAN), over a wireless communications network, or over an asynchronous transfer mode (ATM) network. System 45 includes a first communication terminal 50 for providing information carried by analog voiceband signals, communication link 55 coupled to first communication terminal 50, where the cable may be, for example, electrical or fiber optic, and second communication terminal 65 for receiving the analog voiceband signals from first communication terminal 50 via communication link 55. Second communication terminal 65 comprises filter 60 which converts the analog voiceband signals into digital baseband data. Filter 60 is for example the equalizer described above.

[DESCRIBE THE ADVANTAGES OF THE PRESENT INVENTION.]

Equalizer architecture 10 has several attractive properties. First, the IIR feedback implementation uses a parallel quantization and multiplexing technique, which improves the data throughput, eliminates the A/D converter and simplifies the design. Second, the use of an IIR-model instead of a FIR-model with two-level sequential feedback allows achieving high frequency gain with a simpler structure. Third, all critical circuit components may be operated in two-level or digital signal mode, and their function does not directly rely on the device parasitic parameters, which makes the equalizer performance scales the same with the conventional digital circuits. Consequently, the circuit implementation is reusable and can be directly integrated onto chips.

Since this equalizer is based on the digital-based-analog (DBA) design concept, it will be able to achieve higher design and manufacture efficiency at lower development cost. In one example, an equalizer based on this equalization architecture was developed and a high speed link simulation using the IEEE 1394-1995 Standard, IEEE std. 1394-1995, published Aug. 30, 1996, cable model shows that this equalization method may be used to extend the data speed to the one gigabyte per second (1 Gbt/s) level.

CLAIMS

[ALTHOUGH THE CLAIMS ARE FOUND AFTER THE DETAILED DESCRIPTION, THE FIRST STEP TO WRITING A PATENT APPLICATION IS TO DRAFT THE CLAIMS. THE CLAIMS SHOULD NOT BE TOO NARROW NOR TOO BROAD. YOU SHOULD HAVE A NUMBER OF APPARATUS, METHOD, AND SYSTEM CLAIMS.]

What is claimed is:

[CLAIMS 1, 12, AND 17 ARE INDEPENDENT CLAIMS AND THE REST ARE DEPENDENT CLAIMS.]

1. **An apparatus comprising: a plurality of** quantizers **each configured to compare** a selected threshold signal with an input signal and **generate** an output; a multiplexer, **coupled to** the plurality of quantizers, **that selects** one of the plurality of quantizer outputs according to a frequency response; and a multiplication-accumulation (MAC) unit, **coupled to** the multiplexer, the MAC **to generate** an output based on a previously selected one of the quantizer outputs according to the frequency response.

["WHEREIN" AND "FURTHER COMPRISING" ARE USED THROUGHOUT ALL THE CLAIMS.]

2. **The apparatus of claim 1, wherein** the input signal is an analog signal and a different threshold signal is associated with each quantizer and comprises a value selected to be within the range of the input signal.

3. **The apparatus of claim 2, wherein** the different threshold signals are equally spaced within the range of the input signal.

4. **The apparatus of claim 2, wherein each of the plurality of** quantizers **comprises** a comparator that subtracts the threshold signal from the input signal to generate a representative digital output corresponding to whether the threshold signal is greater than or less than the input signal.

5. **The apparatus of claim 4, wherein** the representative digital output is a high when the threshold signal is less than the input signal.

6. **The apparatus of claim 2, further comprising** a delay line having a plurality of flip flops configured sequentially where each of the plurality of flip flops delays the multiplexer output one clock cycle.

7. **The apparatus of claim 6, wherein** the MAC unit multiplies each delayed output by a coefficient c.sub.n, where "n" is the number of flip flops representative of the frequency response.

8. **The apparatus of claim 7, wherein** the coefficients c.sub.n are determined by the equation:

##EQU3##

where "a.sub.n" represents an order coefficient representative of the frequency response, "n" represents the order of the frequency response and "v.sub.th" the selected threshold signal.

9. **The apparatus of claim 7, wherein** the MAC unit generates an output from the sum of all the delayed outputs multiplied by their respective coefficients, c.sub.n.

10. **The apparatus of claim 9, wherein** the multiplexer selects one of the plurality of quantizer outputs according to the output generated from the MAC unit.

11. **The apparatus of claim 1, wherein** the output sequence generated by the MAC unit based on a previously selected one of the quantizer outputs according to the frequency response, can be represented by the equation:

##EQU4##

[TRANSFORM APPARATUS CLAIM 1 INTO A METHOD CLAIM. FOR EXAMPLE, USE "CONFIGURING TO COMPARE," "SELECTING," AND "GENERATING" INSTEAD OF "COMPARE," "SELECT," AND "GENERATE."]

12. **A method for generating an output from an infinite impulse response (IIR) filter comprising**: **configuring** a plurality of quantizers **to compare** a selected threshold signal with an input signal and generate an output; **selecting** one of a plurality of quantizer outputs according to a frequency response; and **generating** a selection output to select one of the plurality of quantizer outputs based on a previously selected one of the quantizer outputs according to the frequency response.

13. **The method of claim 12, further comprising: configuring** the plurality of quantizers to subtract the threshold signal from the input signal **to generate** a representative digital output corresponding to whether the threshold signal is greater than or less than the input signal.

14. **The method of claim 12, wherein generating** a selection output **comprises** delaying the selected one of the plurality of quantizers outputs, generated from the multiplexer, by at least one cycle.

15. **The method of claim 14, wherein generating** a selection output **comprises multiplying** the delayed output by a coefficient c.sub.n where n is a selected order of the frequency response.

16. **The method of claim 14, wherein generating** a selection output **further comprises** accumulating as a sum all of the delayed outputs multiplied by their respective coefficients.

[TRANSFORM THE APPARATUS CLAIM INTO A SYSTEM CLAIM.]

17. **A communication system comprising**: a first communication terminal **coupled to provide** information carried by an input signal; a communication link **coupled to** the first communication terminal; a second communication terminal **coupled to provide** the input signal from said first communication terminal via said communication link, the second communication terminal **comprising** a filter, which converts the input signal into digital baseband data **comprising**: a plurality of quantizers **each configured to compare** a selected threshold signal with an input signal and generate an output; a multiplexer **that selects** one of the plurality of quantizer outputs according to a frequency response; and a multiplication-accumulation (MAC) unit **that generates** an output based on a previously selected one of the quantizer outputs according to the frequency response.

ABSTRACT

[THE ABSTRACT IS DERIVED FROM CLAIM 1.]

An apparatus including a plurality of quantizers **each configured to compare** a selected threshold signal with an input signal and generate an output, a multiplexer, **coupled to** the plurality of quantizers, **that selects** one of the plurality of quantizer outputs according to a frequency response, and a multiplication-accumulation (MAC) unit, **coupled to** the multiplexer, the MAC **to generate** an output based on a previously selected one of the quantizer outputs according to the frequency response.

DRAWINGS

[INSERT DRAWINGS]

[HERE IS ANOTHER PATENT APPLICATION, RELATED TO THE PRE-VIOUS EXAMPLE. REMEMBER THE BOILERPLATE LANGUAGE IS INDICATED IN BOLD.]

United States Patent   **[PATENT NUMBER]**

[INVENTOR]   **[ISSUE DATE]**

[TITLE]

Description

FIELD OF THE INVENTION

[DESCRIBE THE GENERAL FIELD INVENTION IN WHICH THE PRESENT INVENTION IS LOCATED.]

The invention relates to delay locked loop based circuits for adaptive clock generation.

BACKGROUND

[DESCRIBE THE PRIOR ART AND HOW THE PRESENT INVENTION SOLVES THE PROBLEM OF THE PRIOR ART. THE PROBLEM AND BOILERPLATE LANGUAGE IS INDICATED IN BOLD.]

As the level of integration in semiconductor integrated circuits (ICs) increases, signal delays due to parasitic resistance-capacitance loading become larger. This is especially true of high fan-out global signal lines such as synchronous clocks. Clock signals in modern programmable logic devices may drive several thousand registers. This is a considerable load to the clock driver. Clock tree structures can be implemented on chip to minimize clock skew among registers. However, the base trunk clock driver must be capable of driving this clock tree structure and, as a result, a buffer delay of several nanoseconds is typically incurred.

Circuits using phase locked loop (PLL) are widely used in data communications. An example of such a circuit may be a de-skew clock generation circuit. A typical PLL consists of three on-chip functions and a loop filter. A phase detector measures the phase and frequency difference between an external reference signal and an internal timing signal. Based on the sign and magnitude of the difference, the phase detector drives a charge pump that raises or lowers the voltage level of the loop filter. The loop filter provides a stable voltage input to a voltage-controlled oscillator (VCO). The VCO develops a timing signal that is fed back to the phase detector for comparison with the incoming reference signal. When the reference signal and the VCO timing signal are identical the PLL is "locked" onto the reference signal.

A PLL based circuit may be generally sufficient where power dissipation is not an issue even though communication speeds are high. In certain circuits, communication speeds may range from Megahertz (MHz) to Gigahertz (GHz). In general, however, circuits operating at high speeds are sensitive to power dissipation that results in **overheating of the circuits**. In circuits where power conservation is an issue, power dissipation is also **problematic. As well, problems exist with implementing a PLL in a typical integrated circuit since the PLL uses analog devices such as a phase frequency detector (PFD), charge pump and low pass filter. These problems include, among others, poor stability and performance in a noisy environment.**

SUMMARY

[THE WORDING IS SIMILAR TO THE ABSTRACT, WITH THE ADDITION OF "IN ACCORDANCE WITH AN EMBODIMENT OF THE INVENTION, THERE IS DISCLOSED."]

In accordance with an embodiment of the invention, there is disclosed an apparatus including a phase detector to detect a phase difference between an output clock signal and a local reference clock signal **comprising** a first sampling circuit and a second sampling circuit to cross-sample the output clock signal and the local reference clock signal respectively and a comparator circuit **coupled to** the two sampling circuits that detects the phase difference. A digitally controlled delay line is **coupled to** the output clock signal to adaptively adjust a delay to compensate for the phase difference.

BRIEF DESCRIPTION OF THE DRAWINGS

[BOILERPLATE LANGUAGE IS IN BOLD.]

The accompanying drawings are included to provide a further understanding of the invention, and are incorporated in and constitute a part of this specification. The drawings illustrate embodiments of the invention and, together with the description, serve to explain the principles of the invention. In the drawings,

[DESCRIBE EACH TYPE OF FIGURE AND SPECIFY THAT IT IS "IN ACCORDANCE WITH AN EMBODIMENT OF THE INVENTION." THERE SHOULD BE MANY EMBODIMENTS IN YOUR INVENTION.]

FIG. 1 is a block diagram of a Delay Locked Loop (DLL) based deskew clock generation circuit in accordance with an embodiment of the invention;

FIG. 2 illustrates an output clock that is either aligned or has a constant controllable shift with the input clock in accordance with an embodiment of the invention;

FIG. 3 is a schematic diagram of a phase detector circuit in accordance with an embodiment of the invention;

FIG. 4 is a schematic diagram of a fine digital delay line (FDDL) in accordance with an embodiment of the invention;

FIG. 5 illustrates a typical single-stage differential circuit structure, including a current bias, an input component pair, and a load component pair in accordance with an embodiment of the invention;

FIG. 6 illustrates a symmetric differential complimentary metal-oxide semiconductor (SDCMOS) structure with improved circuit reusability in accordance with an embodiment of the invention;

FIG. 7 illustrates a high speed CMOS differential buffer for either input, intermediate, or output stages in accordance with an embodiment of the invention;

FIG. 8 is a schematic diagram of a coarse digital delay line (CDDL) in accordance with an embodiment of the invention;

FIG. 9 is a schematic diagram of a coarse delay buffer bit which forms part of the CDDL in accordance with an embodiment of the invention;

FIG. 10 is a schematic diagram of a differential multiplexer which forms part of the CDDL in accordance with an embodiment of the invention;

FIG. 11 is a schematic diagram of a differential output buffer structure which forms part of the de-skew clock generation circuit **in accordance with an embodiment of the invention;**

FIG. 12 is a schematic diagram of a system wherein a peripheral controller comprises a de-skew clock generation circuit and is coupled to a processor that is adapted to access data from the peripheral controller **in accordance with an embodiment of the invention.**

DETAILED DESCRIPTION

[YOU SHOULD HAVE MANY EMBODIMENTS THROUGHOUT THE SPECIFICATION. AT THE BEGINNING OF THE DETAILED DESCRIP-TION, DESCRIBE THE BENEFITS OF THE INVENTION. EMPHASIZE THAT VARIOUS EMBODIMENTS WILL BE DESCRIBED TO AID IN THE UNDERSTANDING OF THE INVENTION AND SHOULD NOT BE CON-STRUED AS LIMITATIONS OF THE INVENTION.]

When used in a de-skew clock generation circuit, **in one embodi-ment**, the de-skew clock generation circuit uses a controlled digital delay line to adjust the delay through a pre-defined z-domain algorithm to com-pensate for the phase error. As a result, the output clock will be phase-locked to the input (reference) clock independent of the loading condition. In this manner, a DLL-based de-skew clock generation circuit **achieves very short acquisition time** when compared to the acquisition time of the PLL. Furthermore, the de-skew clock generation circuit is **highly jitter tolerant**. Thus, from above, it can be seen that these features make the DLL de-skew clock generation circuit particularly **suitable for various low power and high-speed applications.**

The operation of a delay locked loop (DLL) may use a voltage con-trolled delay line (VCDL) rather than a VCO to generate the output-timing signal. DLLs **lock onto reference signals faster** than PLLs and they **produce output signals with less jitter**. Multiple chips on a printed cir-cuit board or cores of different sizes within a single system on a chip can experience clock skew. By using DLL technology to shift the phase of the reference clock within each chip or core, **designers can minimize skew and tune a system to perform up to its potential**. DLL devices can be used in each chip or core to **compensate not only for loading differences but also for delays that arise with process, voltage, and temperature (PVT) differences**.

The scheme may be implemented, in one aspect, using a digital-based analog (DBA) design approach, which utilizes analog functions using digital circuits based on certain digital signal processing (DSP) algorithms. **The DBA approach makes the circuit implementation highly scalable and allows the circuit to be directly integrated onto a digital-based chip without degrading its reliability, manufacturability, and testability. Various embodiments will be described to aid in the understanding of the invention and should not be construed as limitations of the invention.**

[REMEMBER TO TELL A STORY WITH THE FIGURES DESCRIBED IN SEQUENCE. IN THIS CASE, FIGURE 1 DESCRIBES THE OVERALL INVENTION. EACH ELEMENT SHOULD BE NUMBERED IN INCREMENTS OF 5, FOR EXAMPLE, DE-SKEW CLOCK GENERATION CIRCUIT 5 AND DCDL 10.]

FIG. 1 is a block diagram of a Delay Locked Loop (DLL) based de-skew clock generation circuit **in accordance with an embodiment of the invention. As shown in the figure**, de-skew clock generation circuit 5 comprises digitally controlled delay line (DCDL) 10, phase detector 15, and output clock buffer 20. **To aid in the understanding of the invention, an overview of the embodiment is given below.**

As shown, in FIG. 2, one purpose of de-skew clock generation circuit 5 is to adaptively adjust the delay so that output clock 22 will be aligned with input clock 21.

In the implemented phase detector 25, **shown in FIG. 3**, input reference clock 21 and output clock 22 rising edges are used to generate two narrow pulses 30 and 35 in order to create a delay. Pulses 30 and 35 are used to control cross sampling of the other (the output and the reference clock) signal. NAND gates 40 and 45 are used as pulse generators for cross-sampling the two signals. That is, inverted input clock and the input clock are sent into NAND gate 40 so that a pulse will be generated for cross-sampling. As well, inverted output clock and output clock 22 are sent into NAND gate 45 so that a pulse will be generated for cross-sampling. As a result, the input clock and the output clock will cross-sample each other when switches 50 and 55 are on. This method is also called differential sampling, which is used, in one aspect, in order to achieve more accuracy.

The sampled signals are then compared using comparator 60 to provide the phase difference of clocks 21 and 22 and to issue the delay line control signals. Comparator 60 will determine if output clock signal 22 is lagging or leading input reference signal 21. At a particular sampling point, if input clock 21 is high and output clock 22 is low, then comparator 60 will detect that output clock signal 22 is lagging input clock 21. If input clock 21 is low and output clock 22 is high then comparator 60 will detect that output clock signal 22 is leading the input. This method of differential sampling eliminates the condition where both signals are high or both are low, **thus increasing accuracy of the circuit**. The information obtained from comparator 60 will then be relayed to fine digital delay line (FDDL) 70 as shown in FIG. 4 and, where necessary, course digital delay line (CDDL) 85.

Digitally controlled delay line (DCDL) 10 unit consists of two sub-blocks 70 and 85 **as shown in FIGS. 4 and 8, respectively**, for either fine or coarse delay compensations. Fine digital delay line (FDDL) 70 uses a plurality of digital controllable differential delay buffer cells 75.

In recent years, there have been significant efforts in the development of mixed-signal circuits, primarily driven by the benefits of cost reduction and performance enhancement through analog and digital circuit integration onto a single chip. Differential circuits, which generally have better signal integrity with larger noise margin and lower noise generation, are widely used in analog and signal-integrity-critical digital circuit implementations of the mixed signal chips. **Shown in FIG. 5** is a single-stage differential circuit structure, consisting of a current bias, an input component pair, and a load component pair. However, these types of circuits generally require very careful selection of all devices and circuit parameters (sizing, biasing, signal swings, gain, speed, drive capability, etc.). Still further, significant tuning or even redesign are usually required for different applications or using different manufacture process technologies due to the highly process dependent nature of the device parameters. Consequently, development of highly reusable differential analog and digital circuits will be very important for the success of the future low cost mixed signal VLSI chips.

In one embodiment of the invention, each buffer cell 75 includes a symmetric differential complimentary metal-oxide semiconductor (SDCMOS) structure. **As shown in FIG. 6, in one embodiment**, the basic SDCMOS circuit uses two CMOS transistor pairs (M1, M2, M3, M4) as the input devices, which extend the input signal to full swing. Additional two CMOS transistor pairs (M5, M6, M7, M8) are used for either current biases or loads. The gates of the bias/load branches are shorted together at points "p" and "n." As can be seen, the entire circuit is symmetric at both left-to-right and top-to-bottom directions. There are three feedback loops in this circuit structure, including the left loop by transistor M1, M2, M5, and M6, the right loop by transistor M3, M4, M7, and M8 and a common mode loop by all transistors as represented by "p" and "n" **in FIG. 6.** For example, a signal at V.sub.in will generate a current I.sub.1 through transistor M6. Likewise, a signal at V.sub.in # will generate a 12 through transistor M8. Both currents will join at common mode "p" to form current I.sub.c, where I.sub.c = I.sub.1 + I.sub.2. In the same manner, a signal at V.sub.in will also generate a current I.sub.3 through transistor M5, and a signal at V.sub.in # will generate a current I.sub.4 through transistor M7. Both currents will join at common mode "n" to form current I.sub.c, where I.sub.c = I.sub.3 + I.sub.4. The circuit configuration illustrated is dynamically self-biased. It can provide higher bias current around the cross-point to achieve zero dc-bias, high speed, and a "soft landing" (avoiding noise and glitches in the signal). These properties generally make SDCMOS circuits very robust on various applications situations (large power supply range, rail-to-rail signal swings, large transistor size range, etc., and very scalable on different manufacture process technologies.

The SDCMOS structure **illustrated in FIG. 6** can be used for various mixed-signal applications. **Shown in FIG. 7** is a high-speed CMOS differential buffer for input, intermediate, or output stages. Two leakage transistors Mp and Mn are used in this circuit to eliminate the dc-path in the down stream circuit by pulling up or down the outputs to rail during the power down mode. In one chip structure, simulation shows that this dc-path elimination technique significantly reduces the static current of the circuit. For a large size down-stream circuit of the same type, such as the interconnect driver/repeater of the clock or critical signals used in one application, this technique can provide significant power reduction.

The total delay that can be adjusted in FDDL 70 can be represented by N (t.sub.2 − t.sub.1) where N is the number of buffer cells in FDDL 70, "t.sub.2" is the total delay needed, and "t.sub.1" is the intrinsic delay due to gates. Time constant (t.sub.2) can be represented by R (C.sub.1 + C.sub.2) where R is the effective resistance, C.sub.1 is the parasitic capacitance, and C.sub.2 is the capacitance controlled by the switch or transistor. Therefore, every time a "1" is input into shift register 80, a fixed delay of "t.sub.2" is generated from the buffer cell **in FIG. 6.** As well, if a "0" is input into shift register array 80, an intrinsic delay of "t.sub.1" is generated from the buffer cell **in FIG. 6.** For example, where N = 6 stages in FDDL and the register contains 1 1 0 0 0 0, then the total delay will be 6 t.sub.1 + 2(t.sub.2 − t.sub.1).

The delay of each cell in DCDL 10 can only be one of two discrete values separated by about 80 pico seconds (ps), controlling through a digital input.

The entire FDDL 70 is controlled through a multi-bit bi-directional shift register array, where a 12-bit register is one embodiment because 12 bits will cover at least a one step delay in CDDL 85. An overflow "O" **as represented in FIG. 4** will occur when the shift register array is filled with "1"s, and another shift-right operation is needed to increase delay. Likewise, an underflow "U" **as represented in FIG. 4** will occur when the shift register array is filled with "0"s, and another shift-left operation is needed to decrease delay.

The number of "0"s or "1"s stored inside the register array can be linearly controlled through the left or right shift operation of the shift registers according to the sign of the phase difference from phase detector 25. For example, if phase detector 25 detects a lag by the sampled output clock, then the delay of the output clock must be decreased, which will be satisfied using a shift left operation. If phase detector 25 detects a lead by the sampled output clock, then the delay of the output clock must be increased, which will be satisfied using a shift right operation.

The coarse delay compensation can be accomplished by CDDL 85 unit **as shown in FIG. 8**, which changes the number of identical delay buffer cells in the clock path using multiplexers 95. **As shown in FIG. 6**, multiplexers 95 are controlled by a multi-bit up/down binary counter 100 which, in turn, is controlled by the underflow (U) or overflow (O) flag signals from FDDL 70 **in FIG. 4**. An exclusive-OR gate is connected to the enable signal, therefore counter 100 will be enabled whenever an underflow (U) or overflow (O) flag signal is sent from FDDL 70. In addition, counter 100 will move up or down depending on whether an overflow bit is detected or not. The total delay will be the sum of delays obtained from FDDL 70 and CDDL 85.

For example, if counter 100 displays 1 0 0 0 0, then a delay of 16 total delay (t.sub.d) will be generated by coarse delay buffer bits **as shown in FIG. 9**, where each bit will yield a fixed delay td since, **in this embodiment**, there is no switching circuitry.

The differential multiplexer (MUX), **as shown in FIG. 10**, will select a delay according to the corresponding bit in up/down counter 100. For example, using the previous example of 1 0 0 0 0, where the bit is "0," the MUX will choose the path where only an intrinsic delay is obtained. And, where the bit is "1," the MUX will choose the path where a delay of 16 t.sub.d is obtained.

Output buffer 20 is used to improve the loading capability of the delay-locked loop circuit. In this embodiment, output buffer 20 consists of four increasingly sized differential buffer stages of similar structure **as shown in FIG. 11**. With output buffer 20, a smaller DCDL 10 can be used. In the de-skew clock application, the feedback clock is usually tapped at the input of the load after output buffer 20. However, early or late output clocks can also be obtained by purposely adding a known delay value in the reference or feedback clock path.

FIG. 12 is a schematic diagram that illustrates a system 105 wherein a peripheral controller 125 comprises a de-skew clock generation circuit 130 similar to the de-skew clock generation circuit described above. **FIG. 12 illustrates** but one application of the invention, that is the personal computer, but may be replaced by other applications such as a workstation, server, Internet driver, or other fabric channels used as a link. **In FIG. 12**, peripheral controller 125 is coupled to processor 115 via a serial or parallel bus 120. Processor 115 is adapted to access data from peripheral controller 125 via bus 120. Memory 110, and display controller 135, may also be coupled to peripheral controller 125 via bus 120. Monitor 140 may also be coupled to display controller 135. Other peripheral devices 145, such as a mouse, CD-ROM and video, may also be coupled to peripheral controller 125.

[DESCRIBE DESIGN ADVANCES OF THE INVENTION.]

Some design advances of the circuit described in this invention include a) high scalability, the feedback control mechanism being based on the pre-designed digital filter algorithm, not the process technology; b) high noise immunity where all critical components are designed using differential circuit technique; c) high reusability and short design time, the design being very modular and regular; d) smaller area and low power, thus there is no explicit capacitor or resistor in the design and most devices in the design are close to minimum. This technology can be used for various de-skew clock generation for either on-chip or off-chip circuits.

[THE FOLLOWING IS BOILERPLATE LANGUAGE THAT CAN BE COPIED INTO YOUR SPECIFICATION.]

In the preceding detailed description, the invention is described with reference to specific embodiments thereof. It will, however, be evident that various modifications and changes may be made thereto without departing from the broader spirit and scope of the invention as set forth in the claims. The specification and drawings are, accordingly, to be regarded in an illustrative rather than a restrictive sense.

CLAIMS

[CLAIMS 1, 14, 19, AND 28 ARE INDEPENDENT CLAIMS; THE REST ARE DEPENDENT CLAIMS.]

What is claimed is:

1. **An apparatus comprising**:

a phase detector to detect a phase difference between an output clock signal and a local reference clock signal, **having** (1) a first sampling circuit and a second sampling circuit to sample the output clock signal by using the reference clock signal, and sample the local reference clock signal by using the output clock signal, respectively and (2) a comparator circuit coupled to the two sampling circuits that detects the phase difference, the first sampling circuit to produce a pulse in response to a transition in the reference clock signal, for sampling the output clock signal; and **a digitally controlled delay line (DCDL) coupled to** the output clock signal **to adaptively adjust** a delay to compensate for the phase difference.

["WHEREIN" AND "FURTHER COMPRISING" IS USED THROUGHOUT ALL THE CLAIMS.]

2. **The apparatus of claim 1, wherein** the digitally controlled delay line (DCDL) **comprises**: a fine digital delay line (FDDL) and a coarse digital delay line (CDDL).

3. **The apparatus of claim 1, further comprising** an output clock buffer coupled to the DCDL **wherein** the output clock buffer **comprises** a plurality of differential buffer stages of similar structure.

4. **The apparatus of claim 2, wherein** the FDDL **further comprises** a plurality of digital controllable differential delay buffer cells, each buffer cell **coupled** to bi-directional shift registers in an array **configured** to perform a shift-left or shift-right operation **according to** the sign of the phase difference.

5. **The apparatus of claim 2, wherein** the CDDL **further comprises** a plurality of differential multiplexers **coupled** to coarse delay buffer bits and to a multi-bit up/down binary counter, and **configured to select** a delay **according to** a corresponding bit in the up/down binary counter.

6. **The apparatus of claim 4, wherein** a number of "0"s and "1"s stored inside the shift register array of the FDDL are linearly controlled

through a shift-left or shift-right operation of the shift registers **according to** the sign of the phase difference from the phase detector unit.

7. **The apparatus of claim 6, wherein** the shift-left operation is performed by inputting a "0" into the right side of the shift register to decrease delay.

8. **The apparatus of claim 7, wherein** the shift-right operation is performed by inputting a "1" into the left side of the shift register to increase delay.

9. **The apparatus of claim 4, wherein** the digital controllable differential delay buffer cell **comprises** a symmetric differential complimentary metal-oxide semiconductor (SDCMOS) circuit.

10. **The apparatus of claim 9, wherein** the SDCMOS circuit **comprises** two CMOS transistor pairs as input devices and two other CMOS transistor pairs used for at least one of current biases and loads.

11. **The apparatus of claim 9 wherein** respective gates of the bias/load transistor pairs are **connected** to differential outputs.

12. **The apparatus of claim 5, wherein** the multiplexers of the CDDL are **controlled** by the multi-bit up/down binary counter which, in turn, is **controlled** by underflow (U) or overflow (O) flag signals from the FDDL.

13. **The apparatus of claim 6, wherein** the shift-left operation is used to decrease the delay and the shift-right operation to increase the delay, so that the output clock is **aligned** with the local reference clock.

[TRANSFORM APPARATUS CLAIM 1 INTO A METHOD CLAIM, FOR EXAMPLE USE "DETECTING" AND "ADJUSTING" INSTEAD OF "DETECT" AND "ADJUST."]

14. **A method** of de-skew clock generation **comprising**:

detecting a phase difference between output and local reference clocks; and

adjusting a delay **to compensate** for the phase differences, **by controlling** a fine digital delay line (FDDL) and a coarse digital delay line (CDDL) **to adjust** the delay, wherein underflow (U) and overflow (O) flag signals from the FDDL **control** a multi-bit up/down binary counter which controls the CDDL.

15. **The method of claim 14, wherein detecting comprises** sampling (1) the output clock by using the local reference clock, and (2) the local

reference clock by using the output clock and **comparing to provide** the phase difference of the two clocks and **to issue** delay line control signals.

16. **The method of claim 14, wherein adjusting** a delay with the FDDL **comprises** linearly **controlling** the number of "0"s or "1"s stored inside a register array through a shift-left or shift-right operation **according to** the sign of the phase difference.

17. **The method of claim 16, wherein shifting left comprises** inputting a "0" into the right side of the register array, in order to decrease delay.

18. **The method of claim 16, wherein shifting right comprises** inputting a "1" into the left side of the register array, in order to increase delay.

[TRANSFORM THE APPARATUS CLAIM INTO A SYSTEM CLAIM. A PERIPHERAL CONTROLLER IS ALWAYS COUPLED TO THE PROCESSOR.]

19. **A system comprising**:

a **processor**; and

a **peripheral controller coupled to the processor**, the processor adapted to access data from the peripheral controller, the peripheral controller comprising a de-skew clock generation circuit comprising:

a phase detector to detect a phase difference between output and local reference clocks, the detector **having** (1) a first sampling circuit and a second sampling circuit to sample the output clock signal by using the reference clock signal, and sample the local reference clock signal by using the output clock signal, respectively, and (2) a comparator circuit coupled to the two sampling circuits that detects the phase difference, the first sampling circuit to produce a pulse in response to a transition in the local reference clock signal, for sampling the output clock signal; and

a digitally controlled delay line (DCDL) to adaptively adjust a delay in the output clock signal to compensate for the phase difference.

20. **The system of claim 19 wherein** the digitally controlled delay line (DCDL) **comprises**:

a fine digital delay line (FDDL) and a coarse digital delay line (CDDL).

21. **The system of claim 19 wherein** the de-skew clock generation circuit **further comprises** an output clock buffer coupled to the DCDL

wherein the output clock buffer **comprises** a plurality of differential buffer stages of similar structure.

22. **The system of claim 20 wherein** the FDDL **further comprises** a plurality of digital controllable differential delay buffer cells, each buffer cell **coupled** to bi-directional shift registers in an array configured **to perform** a shift-left or shift-right operation **according to** the sign of the phase difference.

23. **The system of claim 20 wherein** the CDDL **further comprises** a plurality of differential multiplexers **coupled** to coarse delay buffer bits and to a multi-bit up/down binary counter, and **configured to select** a delay according to a corresponding bit in the up/down binary counter.

24. **The system of claim 22 wherein** the digital controllable differential delay buffer cell **comprises** a symmetric differential complimentary metal-oxide semiconductor (SDCMOS) circuit.

25. **The system of claim 24 wherein** the SDCMOS circuit **comprises** two CMOS transistor pairs as input devices and two other CMOS transistor pairs used for at least one of current biases and loads.

26. **The system of claim 24 wherein** respective gates of the bias/load transistor pairs are **connected** to differential outputs.

27. **The system of claim 23 wherein** the multiplexers of the CDDL **are controlled** by the multi-bit up/down binary counter which, in turn, is controlled by underflow (U) or overflow (O) flag signals from the FDDL.

[THIS IS A VARIATION OF THE OTHER APPARATUS CLAIM. IT IS BETTER TO HAVE A VARIETY, IN CASE ONE CLAIM GETS REJECTED.]

28. **An apparatus comprising**:

a phase detector to detect a phase difference between an output clock signal and a local reference clock signal, **having** (1) a first sampling circuit and a second sampling circuit **to sample** the output clock signal by using the reference clock signal, and **sample** the local reference clock signal by using the output clock signal, respectively, and (2) a comparator circuit **coupled** to the two sampling circuits that detects the phase difference;

a digitally controlled delay line (DCDL) coupled to the output clock signal to adaptively **adjust** a delay to compensate for the phase difference,

wherein the digitally controlled delay line (DCDL) **comprises** a fine digital delay line (FDDL) and a coarse digital delay line (CDDL) and **wherein** the CDDL **further comprises** a plurality of differential multiplexers **coupled** to coarse delay buffer bits and to a multi-bit up/down binary counter, and **configured to select** a delay according to a corresponding bit in the up/down binary counter.

29. **The apparatus of claim 28 wherein** the first sampling circuit is to produce a pulse in response to a transition in the local reference clock signal, for sampling the output clock signal.

30. **The apparatus of claim 28 further comprising** an output clock buffer **coupled to** the DCDL **wherein** the output clock buffer **comprises** a plurality of differential buffer stages of similar structure.

31. **The apparatus of claim 28 wherein** the FDDL **further comprises** a plurality of digital controllable differential delay buffer cells, each buffer cell **coupled to** bi-directional shift registers in an array **configured to perform** a shift-left or shift-right operation **according to** the sign of the phase difference.

32. **The apparatus of claim 31 wherein** the digital controllable differential delay buffer cell **comprises** a symmetric differential complimentary metal-oxide semiconductor (SDCMOS) circuit.

33. **The apparatus of claim 32 wherein** the SDCMOS circuit **comprises** two CMOS transistor pairs as input devices and two other CMOS transistor pairs used for at least one of current biases and loads.

34. **The apparatus of claim 33 wherein** respective gates of the bias/load transistor pairs are **connected** to differential outputs.

35. **The apparatus of claim 28 wherein** the multiplexers of the CDDL are **controlled** by the multi-bit up/down binary counter which is **controlled** by underflow (U) or overflow (O) flag signals from the FDDL.

ABSTRACT

[THE ABSTRACT IS DERIVED FROM CLAIM 1.]

An apparatus including a phase detector to detect a phase difference between an output clock signal and a local reference clock signal comprising a first sampling circuit and a second sampling circuit to cross-sample the output clock signal and the local reference clock signal respectively and a comparator circuit coupled to the two sampling circuits that detects the phase difference.

DRAWINGS

[INSERT DRAWINGS]

Response to Office Action or Amendment

An Office Action is a document written by a patent examiner in response to a patent application after the examiner has examined the application. The Office Action cites prior art and gives reasons why the examiner has allowed, or approved, the applicant's claims and/or rejected the claims.

PATENT   [DOCKET NUMBER]

IN THE UNITED STATES PATENT AND TRADEMARK OFFICE

[SERIAL NUMBER]   [CONFIRMATION NUMBER]

[APPLICANT]   [GROUP ART UNIT]

[FILING DATE]   [EXAMINER]

                                          [CUSTOMER NUMBER]

For: [TITLE]

Via EFS

Commissioner for Patents

P.O. Box 1450

Alexandria, VA 22313-1450

AMENDMENT UNDER 37 C.F.R. Section 1.111

Sir/Madam:

In response to the non-final Office Action dated [INSERT DATE], please amend the above-identified application as indicated below.

Amendments to the claims are reflected in the listing of claims which begins on page 2 of this paper; and

Remarks/Arguments begin on page 9 of this paper.

[APPLICATION NUMBER]   [DOCKET NUMBER]

In Reply to non-final Office Action of [INSERT DATE]

AMENDMENTS TO THE CLAIMS

[THE BOILERPLATE LANGUAGE IS INDICATED IN BOLD.]

This listing of claims will replace all prior versions, and listings, of claims in the application.

Listing of Claims:

[THE BRACKETS AND STRIKETHROUGH ARE USED TO DELETE AN EXPRESSION AND THE UNDERLINES ARE USED TO ADD AN EXPRESSION.]

1. (Currently Amended) A substrate processing apparatus comprising:

    a transport device [[unit]] including a substrate holder holding unit configured to hold the substrate, a rotation driver driving unit configured to rotate the substrate holder holding unit about the center of the substrate holder holding unit as a rotational axis, and a movement driver driving unit configured to move the substrate holder holding unit horizontally;

    a peripheral exposing device [[unit]] including an irradiation device [[unit]] that irradiates light and configured to perform a peripheral exposing process that exposes a peripheral portion of the substrate by irradiating the light to the peripheral portion of the substrate using the irradiation device [[unit]] while rotating the substrate held by the substrate holder holding unit using the rotation driver driving unit;

    a substrate inspecting device [[unit]] including an image pick up device imaging unit that picks up an image and configured to perform a substrate inspecting process that inspects the substrate based on the picked up image while moving the substrate held by the substrate holder holding unit using the movement driver driving unit; and

    a controller control unit configured to control the substrate processing apparatus including the transport device [[unit]], the peripheral exposing device [[unit]], and the substrate inspecting device [[unit]],

    wherein the controller is programmed to control the transport device, the peripheral exposing device and the substrate inspecting device in order to perform a predetermined substrate processing including either one of the peripheral exposing process and the substrate inspecting process or both,

    wherein the control unit controls a predetermined substrate processing to be stopped when the peripheral exposing process is included in the predetermined substrate processing and when it is determined by the controller that an error occurs in the peripheral exposing device [[unit]] and the predetermined substrate processing includes the peripheral exposing process, the controller controls the transport device, the peripheral exposing device and the substrate inspecting device such that the predetermined substrate processing is stopped, and

when it is determined by the controller that an error occurs in the substrate inspecting device and the predetermined substrate processing includes the substrate inspecting process while no error occurs in both of the transport device and the peripheral exposing device, the controller controls the transport device, the peripheral exposing device and the substrate inspecting device such that the control unit controls the substrate inspecting process to be skipped while the peripheral exposing process continues when no error occurs in both of the peripheral exposing unit and the transport unit, the substrate inspecting process is included in the predetermined substrate processing, and the error occurs in the substrate inspecting unit.

2. (Currently Amended) The substrate processing apparatus of claim 1, wherein the peripheral exposing device [[unit]] includes a light emitting device [[unit]] configured to emit the light to be irradiated by the irradiation device [[unit]], and

the controller control unit controls the transport device, the peripheral exposing device and the substrate inspecting device such that the predetermined substrate processing to be is stopped when the peripheral exposing process is included in the predetermined substrate processing and an error occurs in the light emitting device [[unit]].

3. (Currently Amended) The substrate processing apparatus of claim 1, wherein the controller control unit controls the transport device, the peripheral exposing device and the substrate inspecting device such that the predetermined substrate processing to be is stopped when the peripheral exposing process is included in the predetermined substrate processing and the peripheral exposing device [[unit]] is in an excessive temperature rise a state where temperature is rising.

4. (Currently Amended) The substrate processing apparatus of claim 1, wherein the peripheral exposing device [[unit]] includes

a shutter [[unit]] configured to transmit/interrupt the light to be irradiated by the irradiation device [[unit]], and

the control unit controls <u>the transport device, the peripheral exposing device and the substrate inspecting device such</u> that the predetermined substrate processing <s>to be</s> <u>is</u> stopped when the peripheral exposing process is included in the predetermined substrate processing and an error occurs in the shutter [[unit]].

5. (Currently Amended) The substrate processing apparatus of claim 1, wherein the substrate inspecting <u>device</u> [[unit]] includes an image <u>processor</u> <s>processing unit</s> configured to process <u>an</u> image picked up by the <u>image pick up device</u> <s>imaging unit</s>, and

the <u>controller</u> <s>control unit</s> controls <u>the transport device, the peripheral exposing device, and the substrate inspecting device such</u> that the substrate inspecting process <s>to be</s> <u>is</u> skipped when no error occurs is both of the peripheral exposing <u>device</u> [[unit]] and the transport <u>device</u> [[unit]], the substrate inspecting process is included in the predetermined substrate processing, and a communication error occurs between the <u>image pick up device</u> <s>imaging unit</s> and the image <u>processor</u> <s>processing unit</s>.

6. (Currently Amended) The substrate processing apparatus of claim 1, wherein the transport <u>device</u> [[unit]] includes an alignment <u>device</u> [[unit]] configured to align a rotation angular position of the substrate held by the substrate <u>holder</u> <s>holding unit</s>.

7. (Currently Amended) A substrate processing method for performing a predetermined substrate processing <s>comprising at least one of</s> <u>using a substrate processing apparatus including a transport device, a peripheral exposing device, a substrate inspecting device, and a controller, the method comprising</u>:

<s>a peripheral exposing process configured to expose</s> <u>exposing</u> a peripheral portion of a substrate by irradiating light to the peripheral portion of the substrate using an irradiation <u>device</u> <s>unit of a</s> <u>included in the</u> peripheral exposing <u>device</u> [[unit]] while rotating the substrate held by a substrate <u>holder included in the transport device</u> <s>holding unit</s> about the center of the substrate <u>holder</u> <s>holding</s> unit as a rotational axis; [[and]]

<s>a substrate inspecting process configured to pick</s> <u>picking</u> up an image of the substrate using an <u>image pick up device</u> <s>imaging</s>

~~unit of a~~ included in the substrate inspecting device [[unit]] and ~~inspect~~ inspecting the substrate based on the picked up image while moving the substrate held by the substrate holder ~~holding unit~~ horizontally;[[,]]

performing a predetermined substrate processing including either one of the peripheral exposing process and the substrate inspecting process or both by controlling the transport device, the peripheral exposing device and the substrate inspecting device;

~~wherein the substrate processing method further comprising~~ when it is determined by the controller that an error occurs in the peripheral exposing device and the predetermined substrate processing includes the peripheral exposing process, stopping the predetermined substrate processing; ~~when the peripheral exposing process is included in the predetermined substrate processing and an error occurs in the peripheral exposing unit~~, and

when it is determined by the controller that an error occurs in the substrate inspecting device and the predetermined substrate processing includes the substrate inspecting process while no error occurs in both of the transport device and the peripheral exposing device, skipping the ~~substrate~~ picking up and the inspecting ~~process~~ while continuing the peripheral exposing process ~~when no error occurs in both of the peripheral exposing unit and a transport unit that holds the substrate using the substrate holding unit and rotates and horizontally moves the substrate, the substrate inspecting process is included in the predetermined substrate processing, and an error occurs in the substrate inspecting unit.~~

8. (Currently Amended) The substrate processing method of claim 7, wherein the peripheral exposing device [[unit]] includes a light emitting device [[unit]] configured to emit the light to be irradiated by the irradiation device [[unit]] and

the method further comprising stopping the predetermined substrate processing when the ~~peripheral~~ exposing ~~process~~ is included in the predetermined substrate processing and an error occurs in the light emitting device [[unit]].

9. (Currently Amended) The substrate processing method of claim 7, further comprising stopping the predetermined substrate

processing when the ~~peripheral~~ exposing ~~process~~ is included in the predetermined substrate processing and the peripheral exposing <u>device</u> [[unit]] is in ~~an excessive temperature rise~~ <u>a</u> state <u>where temperature is rising</u>.

10. (Currently Amended) The substrate processing method of claim 7, wherein the peripheral exposing <u>device</u> [[unit]] includes a shutter [[unit]] configured to transmit/interrupt the light to be irradiated by the irradiation <u>device</u> [[unit]], and

   the method further comprising stopping the predetermined substrate processing when the ~~peripheral~~ exposing ~~process~~ is included in the predetermined substrate processing and an error occurs in the shutter [[unit]].

11. (Currently Amended) The substrate processing method of claim 7, wherein the substrate inspecting <u>device</u> [[unit]] includes an image <u>processor</u> ~~processing unit~~ configured to process the image picked up by the <u>image pick up device</u> ~~imaging unit~~, and

   the method further comprising skipping the ~~substrate~~ <u>picking up and the</u> inspecting ~~process~~ when no error occurs in both of the peripheral exposing <u>device</u> [[unit]] and the transport <u>device</u> [[unit]], the substrate <u>picking up and the</u> inspecting ~~process is~~ <u>are</u> included in the predetermined substrate processing, and a communication error occurs between the <u>image pick up device</u> ~~imaging unit~~ and the image <u>processor</u> ~~processing unit~~.

12. (Currently Amended) The substrate processing method of claim 7, wherein an alignment process of aligning a rotation angular position of the substrate held by the substrate <u>holder</u> ~~holding unit~~ is performed before the predetermined substrate processing is performed.

13. (Original) A non-transitory computer readable recording medium storing a computer executable program that, when executed, causes a computer to perform the substrate processing method of claim 7.

REMARKS

[THE BOILERPLATE LANGUAGE IS INDICATED IN BOLD. WE WILL EXAMINE THE MOST COMMON GROUNDS FOR REJECTION IN A RESPONSE TO OFFICE ACTION.]

Applicants respectfully requests reconsideration of this application in view of the foregoing amendments and following remarks.

[ALWAYS RECITE THE STATUS OF THE CLAIMS AT THE BEGINNING OF YOUR REMARKS.]

Status of the Claims
Claims 1–13 are pending in this application, and stand rejected. By this amendment, claims 1–12 are amended. No new matter has been added by this amendment.

Interpretation under 35 U.S.C. Section 112, 6th paragraph
The Office Action indicates that:

Claim limitation "transport unit; substrate holding unit; driving unit; exposing unit; irradiating unit; inspecting unit; control unit, shutter unit; emitting unit; processing unit; imaging unit," has/have been interpreted under 35 U.S.C. Section 112(f) or pre-AIA 35 U.S.C. Section 112, sixth paragraph, because it uses/they use a generic placeholder "unit" coupled with functional language without reciting sufficient structure to achieve the function. Furthermore, the generic placeholder is not preceded by a structural modifier.

First of all, Applicants note that the cited phrases such as, for example, "transport unit; substrate holding unit; driving unit; exposing unit; irradiating unit; inspecting unit; control unit; shutter unit; emitting unit; processing unit; imaging unit," **do NOT explicitly use the phrase "means for" or "step for" along with functional language, and thus do NOT invoke 35 U.S.C. Section 112, sixth paragraph. See, for example, MPEP 2181 (I).**

Moreover, Applicants believe that each of the cited phrases is a structural term and is not a mere substitute for the term "means for." For example, the phrase "a transport unit" **is a structural term and different from non-structural terms such as "mechanism for," "module for," or "system for," as specifically listed in MPEP 2181 (I)(A).**

Additionally, Applicants believe that these phrases have been well recognized by one ordinary skill in the art as the name of structures without the need to resort to other portions of the specification. For example, the relevant portion of the MPEP states that "35 U.S.C. 112, paragraph 6 will not apply if persons of ordinary skill in the art read-

ing the specification would understand the term to be the name of the structure that performs the function, even when the term covers a broad class of structure or identifies the structures by their function."

Nonetheless, claims have been amended for further clarification. In particular, amended claim 1 recites, inter alia, "a transport device . . .," "a substrate holder . . .," "a rotation driver . . .," "a movement driver . . .," "a peripheral exposing device . . .," "irradiation device . . .," "a substrate inspecting device . . .," "an image pick up device . . .," and "a controller . . . ."

Applicants believe that the amended phrases are structural terms and well recognized by one ordinary skill in the art as the name of structures without the need to resort to other portions of the specification, i.e., should not invoke 35 U.S.C. Section 112, sixth paragraph.

Rejections under 35 U.S.C. Section 112
[STATE THE REASONS FOR REJECTION.]

In paragraph two (2) of the Office Action, claims 1–13 have been rejected under 35 U.S.C. Section 112, second paragraph, as allegedly being indefinite. The Office Action indicates, inter alia, that the phrases "to be stopped when" in claim 1 is grammatically connected to three separate and contradictory clauses. The Office Action further indicates that the term "excessive" in claim 3 is a relative term.

In response, claims 1 and 3 have been amended as indicated above addressing the rejections under this category.

Reconsideration and withdrawal of the rejection of claims 1–13 under 35 U.S.C. Section 112, second paragraph, is respectfully requested.

Rejections under 35 U.S.C. Section 103
[THIS INFORMATION CAN BE FOUND IN THE INTRODUCTION OF THE OFFICE ACTION. THAT IS, STATE ALL THE REJECTIONS THAT ARE CITED AGAINST YOUR CLAIMS.]

In paragraph 11 of the Office Action, claims 1, 2, 4, 6–8, 10, and 12 have been rejected under 35 U.S.C. 103(a) as allegedly being unpatentable over [INSERT PUBLICATION NUMBER] to [INSERT REFERENCE1] and further in view of [INSERT PUBLICATION

NUMBER] to [INSERT REFERENCE2] and [INSERT PUBLICATION NUMBER] to [INSERT REFERENCE3]. In paragraph 12 of the Office Action, claims 3 and 9 have been rejected under 35 U.S.C. Section 103(a) as allegedly being unpatentable over [INSERT REFERENCE1] and further in view of [INSERT REFERENCE2] and [INSERT REFERENCE3] as applied to claims 1 and 7 above, and further in view of [INSERT PUBLICATION NUMBER] to [INSERT REFERENCE4]. In paragraph 14 of the Office Action, claims 5 and 11 have been rejected under 35 U.S.C. Section 103(a) as allegedly being unpatentable over [INSERT REFERENCE1] and further in view of [INSERT REFERENCE2] and [INSERT REFERENCE3] as applied to claims 1 and 7 above, and further in view of [INSERT PUBLICATION NUMBER] to [INSERT REFERENCE5].

[DESCRIBE IN DETAIL WHY THE CLAIMS ARE REJECTED BY CITING THE PRIMARY AND SECONDARY REFERENCES.]

In rejecting claim 1, in addition to the primary reference [INSERT REFERENCE1], the Office Action indicates, inter alia, that the secondary reference, [INSERT REFERENCE2], discloses that the control unit controls the predetermined substrate processing to be skipped **citing paragraph [0059] of [INSERT REFERENCE2]. The Office Action also indicates that another secondary reference, [INSERT REFERENCE3], discloses** that the control unit controls the substrate inspecting process to be skipped citing paragraph [0084] of **[INSERT REFERENCE3]**.

One of the aspects of the substrate processing apparatus of claim 1 includes, inter alia, a controller configured to control the substrate processing apparatus including the transport device, the peripheral exposing device, and the substrate inspecting device. In particular, the controller is programmed to control these devices to perform a predetermined substrate processing including either one of the peripheral exposing process and the substrate inspecting process or both. When it is determined by the controller that an error occurs in the peripheral exposing device and the predetermined substrate processing includes the peripheral exposing process, the controller controls the devices such that the predetermined substrate processing stopped. When it is determined by the controller that an error occurs in the substrate inspecting device and the predetermined substrate processing includes the substrate inspecting process while no error occurs in both of the transport device and the peripheral exposing device, the controller controls the device such that the substrate inspecting process is skipped while the peripheral exposing process continues.

[EXPLAIN THAT CLAIM 1 NOW OVERCOMES THE REJECTION.]

Claim 1 has been amended to further clarify these aspects of invention. Support for the amendment may be found, for example, in paragraphs [0118]–[0130] **of the specification as originally filed. Another independent claim 7 has been amended in a similar manner to claim 1 as discussed above**.

[INSERT REFERENCE2] discloses a scanning exposure apparatus in which the exposing process continues even after the control system determines that the condition is out of the tolerance. **Applicants note that the cited portion of [INSERT REFERENCE2] (i.e., paragraph [0059]) discloses** that the so-called "forced exposure" should be limited within a finite number of successive shots, and the job stops when the number of forced exposure exceeds the finite number.

[INSERT REFERENCE3] discloses an image pick up apparatus that detects and notifies the states where degradation of the image quality occurs due to heat. **Applicants note that the cited portion of [INSERT REFERENCE3] (i.e., paragraph [0084]) discloses** that some of the processes are skipped when the temperature of the image pick up device exceeds the permissible temperature.

First of all, each of [INSERT REFERENCE2] and [INSERT REFERENCE3] discloses an exposure process and an inspecting process independently, and **substantially different from the inventive aspects of claim 1 as amended described above** where the exposure process and the inspecting process are selectively controlled in an integrated system including both of the peripheral exposure device and the inspecting device.

[EXPLAIN THAT THERE IS NO MOTIVATION OR SUGGESTION TO COMBINE THE REFERENCES.]

Moreover, as Applicants understand it, there is no motivation to combine these two references in each of [INSERT REFERENCE2] and [INSERT REFERENCE3].

Secondly, the control conditions in the references (e.g., the number of forced exposure **in [INSERT REFERENCE2]** and the temperature in the image pick up device) **are substantially different from claim 1 as amended. For example, amended claim 1 recites, inter alia, that** "when it is determined by the controller that an error occurs in the peripheral exposing device and the predetermined substrate processing includes the peripheral exposing process," and "when it is determined by the controller that an error occurs in the substrate inspecting device and the predetermined substrate processing includes the substrate inspecting process."

Another secondary reference, [INSERT REFERENCE4], is cited as disclosing that the peripheral exposing unit is in an excessive temperature rise state. **[INSERT REFERENCE5] is cited as disclosing** a communication error occurring between the imaging unit and the image-processing unit. **However, none of these secondary references shows or suggests the inventive aspects of claim 1 as amended discussed above**.

[ONLY CLAIMS 1 AND 7 NEED TO BE ADDRESSED SINCE THEY ARE BOTH INDEPENDENT CLAIMS.]

Accordingly, claims 1 and 7 as amended are believed not rendered obvious in view of the references cited by the Examiner (i.e., INSERT REFERENCES 1, 2, 3, 4, AND 5), either taken alone or in combination, for at least the reasons discussed above.

Reconsideration and withdrawal of the rejection of claims 1 and 7 under 35 U.S.C. Section 103(a) is respectfully requested.

[EXPLAIN THAT YOU HAVE NOT ADDRESSED THE REJECTIONS OF THE DEPENDENT CLAIMS BUT RESERVE THE RIGHT TO DO SO IN THE FUTURE.]

Applicants have chosen in the interest of expediting prosecution of this patent application to distinguish the cited documents from the pending claims as set forth above. However, these statements should not be regarded in any way as admissions that the cited documents are, in fact, prior art. Also, Applicants have not individually addressed the rejections of the dependent claims because Applicants submit that independent claims 1 and 7 from which they respectively depend are in condition for allowance as set forth above. Applicants however reserve the right to address such rejections of the dependent claims should such be necessary.

Applicants believe that the application as amended is in condition for allowance and such action is respectfully requested.

United States Patent and Trademark Office

AMENDMENT

[INSERT DATE]

Commissioner for Patents

P.O. Box 1450

Alexandria, VA 22313-1450

Sir/Madam:

Applicant submits the following amendments and remarks in response to final Office Action mailed [INSERT DATE], and in support of the request for continued examination (RCE) filed concurrently herewith.

IN THE CLAIMS

This Listing of Claims will replace all prior versions, and listings, of claims in the application:

Listing of Claims:

1. (withdrawn) A method of adjusting at least one working unit of a harvesting machine, comprising the steps of

taking reference pictures or reference picture series of a crop-material flow at a position downstream from a particular working unit in a crop-material conveyance path at various setting states with certain control parameter setting values of the working unit;

storing the reference pictures or reference picture series such that they are assigned to control parameter setting values belonging to a particular setting state;

taking working pictures or working picture series of a working crop material flow for analysis;

selecting a reference picture or reference picture series that is most representative of said working crop material flow based on an analysis and comparison of the working crop-material flow depicted in the working picture or working picture series with the reference picture or reference picture series; and

adjusting the particular working unit using the control parameter setting values assigned to the selected reference picture or the selected reference picture series.

2. (withdrawn) A method as defined in claim 1, further comprising within an optimization procedure, intentionally varying at least one control parameter or a group of control parameters of the working unit while holding remaining control parameters constant; and producing a picture or a picture series at certain settings of this control parameter or the group of control parameters.

3. (withdrawn) A method as defined in claim 1, further comprising assigning the control parameters setting values of other working units of the harvesting machine that existed at a point in time when the pictures were taken, to the pictures or picture series; and storing information about harvesting conditions that existed at the point in time when the pictures were taken.

4. (withdrawn) A method as defined in claim 1, further comprising not taking a picture or a picture series of the crop-material flow at a certain adjustment state until a certain period of time has expired, after a particular setting state of the working unit was implemented.

5. (withdrawn) A method as defined in claim 1, further comprising displaying the pictures or picture series taken at various setting states to an operator of the harvesting machine for selection.

6. (withdrawn) A method as defined in claim 5, wherein said displaying includes providing the display of the pictures or picture series belonging to the various setting states at least partially and parallel.

7. (withdrawn) A method as defined in claim 1, further comprising automatically analyzing the pictures or picture series taken at various setting states; and, based on an analysis result, selecting a picture or picture series.

8. (withdrawn) A method as defined in claim 7, wherein said selecting includes a selection of a picture or picture series automatically based on the analysis result.

9. (withdrawn) A method as defined in claim 1, further comprising using reference pictures to analyze a picture series.

10. (withdrawn) A method as defined in claim 9, further comprising assigning quality information related to the crop-material flow depicted in a particular reference picture, to the reference pictures.

11. (withdrawn) A method as defined in claim 1, further comprising taking the picture or picture series in a crop-material conveyance path between an outlet of a cleaning device and a crop-material storage device or a crop-material outlet of the harvesting machine.

12. (withdrawn) A method as defined in claim 1, further comprising taking pictures or picture series at a setting state, at various positions in the crop-material conveyance path.

13. (withdrawn) A harvesting machine, comprising:

a working unit;

a picture detector located at a position downstream from the working unit in a crop-material conveyance path to take working pictures or working picture series of a crop-material flow;

a control unit configured such that said working unit is brought into various setting states via a control with certain control parameter setting values, and reference pictures or reference picture series of the crop-material flow are taken at various setting states of said working unit; and

a memory device for storing the reference pictures or reference picture series such that they are assigned to control parameter setting values belonging to a particular setting state;

a selection unit for selecting a reference picture or a reference picture series based on an analysis and comparison of the crop-material flow depicted in the working picture or working picture series with the reference picture or reference picture series;

wherein said control unit is configured such that the working unit is adjusted using the control parameter setting values assigned to the selected reference picture or the selected reference picture series.

14. (withdrawn) A harvesting machine as defined in claim 13, further comprising an analysis unit for automatically analyzing the picture or picture series taken at various setting states.

15. (withdrawn) A harvesting machine as defined in claim 14, wherein said selection unit is configured such that a selection of a picture or a picture series takes place automatically based on said analysis.

16. (withdrawn) A harvesting machine as defined in claim 13, wherein said selection unit includes a display device to display the pictures or picture series taken at various setting states to an operator of the harvesting machine for selection, and an acquisition device for capturing a selection command from an operator.

17. (withdrawn) A harvesting machine as defined in claim 13, further comprising several image detectors located at various positions of the crop-material conveyance path.

18. (currently amended) A processor controlled method of adjusting at least one working unit of a harvesting machine, comprising the steps of

utilizing a control unit processor to take ~~taking~~ reference pictures or reference picture series of a crop-material flow at a position downstream from a particular working unit in a crop-material conveyance path at various setting states with certain control parameter setting values of the working unit;

utilizing a control unit processor to store ~~storing~~ the reference pictures or reference picture series while assigning ~~in cooperation with a control unit processor to assign~~ the picture or picture series to control parameter setting values belonging to a particular setting state of the working unit;

utilizing a control unit processor to select ~~selecting~~ a reference picture or reference picture series based on a qualitative analysis of the crop material flow depicted in the reference picture or reference picture series implemented by the control unit processor; and

adjusting the working unit using the control parameter setting values assigned to the selected reference picture or the selected reference picture series by ~~in cooperation with~~ the control unit processor.

19. (previously presented) The method as defined in claim 18, further comprising within an optimization procedure implemented by the control unit processor, intentionally varying at least one control parameter or a group of control parameters of the working unit while holding remaining control parameters constant; and producing a picture or a picture series at certain settings of this control parameter or the group of control parameters.

20. (previously presented) The method as defined in claim 18, further comprising the control unit processor: assigning the control parameters setting values of other working units of the harvesting machine that existed at a point in time when the pictures were taken, to the pictures or picture series; and storing information about harvesting conditions that existed at the point in time when the pictures were taken.

21. (previously presented) The method as defined in claim 18, further comprising the control unit processor: not taking a picture or a picture series of the crop-material flow at a certain adjustment state until a certain period of time has expired, after a particular setting state of the working unit was implemented.

22. (previously presented) The method as defined in claim 18, further comprising the control unit processor: displaying the pictures or picture series taken at various setting states to an operator of the harvesting machine for selection.

23. (previously presented) The method as defined in claim 22, wherein said displaying includes providing the display of the pictures or picture series belonging to the various setting states at least partially and parallel.

24. (previously presented) The method as defined in claim 18, further comprising the control unit processor: automatically analyzing the pictures or picture series taken at various setting states; and, based on an analysis result, selecting a picture or picture series.

25. (previously presented) The method as defined in claim 24, wherein said selecting includes a selection of a picture or picture series automatically based on the analysis result.

26. (previously presented) The method as defined in claim 18, further comprising the control unit processor: using reference pictures to analyze a picture series.

27. (previously presented) The method as defined in claim 26, further comprising the control unit processor: assigning quality information related to the crop-material flow depicted in a particular reference picture, to the reference pictures.

28. (previously presented) The method as defined in claim 18, further comprising the control unit processor: taking the picture or picture series in a crop-material conveyance path between an outlet of a cleaning

device and a crop-material storage device or a crop-material outlet of the harvesting machine.

29. (previously presented) The method as defined in claim 18, further comprising the control unit processor: taking pictures or picture series at a setting state, at various positions in the crop-material conveyance path.

30. (currently amended) A harvesting machine, comprising:

a working unit;

a picture detector located at a position downstream from the working unit in a crop-material conveyance path to take reference pictures or reference picture series of a crop-material flow;

a control unit including a control unit processor <u>programmed</u> ~~configured~~ to automatically bring said working unit into various setting states via a control with certain control parameter setting values and to control taking reference pictures or reference picture series of the crop-material flow at various setting states of said working unit, the control unit comprising:

a memory device for storing the reference pictures or reference picture series such that they are assigned to control parameter setting values belonging to a particular setting state of the working unit; and

a selection unit for selecting a reference picture or a reference picture series based on a qualitative analysis of crop-material flow depicted in the reference picture or reference picture series;

wherein said control unit processor coordinates transfer of pictures or picture series taken by the picture detector into the memory device, controls selecting by the selecting unit and controls adjustment of the working unit using the control parameter setting values assigned to the selected reference picture or the selected reference picture series in response to selecting by the selection unit.

31. (previously presented) The harvesting machine as defined in claim 30, wherein the control unit further comprises an analysis unit for automatically analyzing the picture or picture series taken at various setting states.

32. (currently amended) The harvesting machine as defined in claim 31, wherein said selection unit is <u>programmed to operate</u> ~~configured~~ such that a selection of a picture or a picture series takes place automatically based on said analysis.

33. (previously presented) The harvesting machine as defined in claim 30, further comprising a display device connected to and controlled by the selection unit to display the pictures or picture series taken at various setting states to an operator of the harvesting machine for selection, and an acquisition device for capturing a selection command from an operator.

34. (previously presented) The harvesting machine as defined in claim 30, further comprising several image detectors located at various positions of the crop-material conveyance path.

REMARKS

[DESCRIBE THE REJECTED CLAIMS AS MENTIONED IN THE OFFICE ACTION.]

The present Amendment is submitted in response to the final Office Action mailed on [INSERT DATE].

The final Office Action allows harvesting machine claims 30–34 (but objects to claims 30 and 32 on formalities), indicates allowability of method claims 24 and 25, again rejects method claims 18–29 under both Section 101 and U.S.C. Section 112, first paragraph and rejects claims 18–23 and 26–29 under Section 102(b) over [INSERT REFERENCE].

Applicant expresses his gratitude to the Examiner for the allowance of claims 30–34. For that matter, applicant has amended claims 30 and 32 to utilize "programmed to" instead of "configured to" in order to clearly indicate and reflect the programmed-controlled operation of the control unit processor and selection unit, as suggested. **Applicant respectfully requests withdrawal of the objection to claims 30 and 32, therefore.**

Applicant further expresses his gratitude for the indication of the allowability claims 24 and 25. Applicant, however, opts not to amend claims 24 and 25 to include the limitations of claim 18 because of his strong belief in the patentability of claim 18 over [INSERT REFERENCE] in view instant amendments to claim 18 (as shown above), and the arguments below.

[THE EXAMINER INDICATES THAT THE CLAIMS PROVIDE NO UTILITY BECAUSE EACH STEP OF THE METHOD CLAIM COULD BE PERFORMED BY HAND OR ON A COMPUTER.]

To support the rejection of claims 18–29 under Section 101, the Examiner asserts that the claims fail to explicitly recite a machine necessary to execute the process steps, that *each* step of the method could be performed by hand or on a computer.

To support the rejection under Section 112, first paragraph, the Examiner asserts that the claims provide no utility because *each* step of the method could be performed by hand or on a computer.

The Examiner summarizes her position by stating that "[t]here is no structure in the claims that require that the method be executed automatically, which applicant appears to be arguing."

[THE APPLICANT AMENDS THE CLAIM IN ORDER TO EXPLICITLY RECITE THE "MACHINE."]

In response to the rejection of claims 18–29 under both Section 101 and U.S.C. Section 112, first paragraph, applicants amend claim 18 to explicitly recite the "machine" a machine necessary to execute the process steps thereby making clear that the invention as claimed is meant to be executed by machine, automatically, and not optionally by hand.

That is, claim 18 as amended now calls out a processor controlled method of adjusting at least one working unit of a harvesting machine, comprising

utilizing a control unit processor to take reference pictures or reference picture series of a crop-material flow at a position downstream from a particular working unit in a crop-material conveyance path at various setting states with certain control parameter setting values of the working unit;

utilizing a control unit processor to store the reference pictures or reference picture series while assigning the picture or picture series to control parameter setting values belonging to a particular setting state of the working unit;

utilizing a control unit processor to select a reference picture or reference picture series based on a qualitative analysis of the crop material flow depicted in the reference picture or reference picture series implemented by the control unit processor; and

adjusting the working unit using the control parameter setting values assigned to the selected reference picture or the selected reference picture series by the control unit processor.

[THE INVENTION IS NOW UNDERSTOOD TO BE SUPPORTED BY A SPECIFIC AND SUBSTANTIAL ASSERTED UTILITY OR A WELL-ESTABLISHED UTILITY.]

In view of these added limitations to claim 18, and in view of the fact that the processor controlled method for adjusting a working unit of a harvesting machine as amended now explicitly recites the machine (i.e., the control unit processor) necessary to execute each of the steps of the process, automatically, the invention as claimed is understood to be supported by either a specific and substantial asserted utility or a well-established utility, and the skilled artisan would readily understand how to use the claimed invention.

[APPLICANT REQUESTS WITHDRAWAL OF THE REJECTION.]

Applicant, therefore, respectfully requests withdrawal of the rejection of claims 18–29 under 35 U.S.C. Section 101, and the rejection of claims 18–29 under 35 U.S.C. Section 112, first paragraph.

In response to the substantive rejection of the claims 18–34 under 35 U.S.C. Section 102(b) over [INSERT REFERENCE], applicant respectfully asserts that independent claim 18 as amended now makes clear that the control unit processor takes the reference pictures at various setting states, etc., memory stores the reference pictures while assigning them to control parameter setting values belonging to a particular setting state of the working unit, selects a reference picture based on a qualitative analysis of the crop material flow depicted in the reference picture and adjusts the working unit using the control parameter setting values assigned to the selected reference picture.

[INSERT REFERENCE] does not disclose utilizing a processor to store reference pictures to assign them to control parameter setting values belonging to a particular setting state of the working unit or to select a reference picture based on a qualitative analysis of the crop-material flow depicted in the reference picture, as claimed.

[THE REFERENCE DOES NOT ANTICIPATE THE CLAIM.]

In view of the fact that independent claim 18 as amended now recites these limitations, which [INSERT REFERENCE] does not, [INSERT REFERENCE] does not anticipate independent claim 18. Hence, claim 18 and claims 19–29 that depend therefrom are patentable under Section 102(b) over [INSERT REFERENCE], and applicant respectfully requests withdrawal of the rejections thereunder.

[THE APPLICATION AS AMENDED IS IN CONDITION FOR ALLOWANCE.]

Accordingly, the application as amended is believed to be in condition for allowance. Action to this end is courteously solicited. However, should the Examiner have any further comments or suggestions, the undersigned would very much welcome a telephone call in order to discuss appropriate claim language that will place the application in condition for allowance.

UNITED STATES PATENT AND TRADEMARK OFFICE

AMENDMENT

[INSERT DATE]

Commissioner for Patents

P O Box 1450

Alexandria, VA 22313-1450

Sir/Madam:

Responsive to the Office Action mailed [INSERT DATE], applicants request entry of the following amendments and remarks.

IN THE CLAIMS

This Listing of Claims will replace all prior versions, and listings, of claims in the application:

Listing of Claims:

1. (currently amended) A route planning system for agricultural working machines, comprising means for assigning a defined working width to the agricultural working machines to generate driving routes in a territory, and for dynamic adaptation of the planned driving route, including automatically generating a new driving path for an agricultural working machine in response to an operator changing the machine's working sequence for the territory at any time, wherein the new driving path is automatically worked by the working machine thereby ensuring that the driving route to be covered is flexibly adaptable to changing external conditions but for such as driving around obstacles in working machine paths, thereby largely relieving the operator of the agricultural working machine of the task of performing laborious steering maneuvers.

2. (withdrawn) The route planning system for agricultural working machines as defined in claim 1, wherein said means is formed so that the planned driving route (14, 14') is adapted dynamically as a function of the actual machine position (31) and the actual machine orientation (33).

3. (withdrawn) The route planning system for agricultural working machines as defined in claim 1, wherein said means is formed so

that the dynamic adaptation of the driving route (14, 14') is carried out permanently.

4. (withdrawn) The route planning system for agricultural working machines as defined in claim 1, wherein said means is formed so that the driving route (14, 14') is generated based on a large number of driving paths (25, 26), and the driving paths (25, 26) are determined based on optimization criteria (11).

5. (withdrawn) The route planning system for agricultural working machines as defined in claim 1, wherein said means is formed so that the next driving path (25, 26) to be worked is selected based on optimization criteria (11).

6. (withdrawn) The route planning system for agricultural working machines as defined in claim 5, wherein said means is formed so that the optimization criteria (11) can be "shortest driving route/processing time," "small proportion of unproductive auxiliary time," "short auxiliary drives between successive driving paths (25, 26) to be worked," "recognition and working of known driving routes (14, 14') and sequences," "short turn-around routes (36)," and "minimize routes between agricultural working machine (3, 4) and hauling vehicle (35)."

7. (original) The route planning system for agricultural working machines as defined in claim 1, wherein said means is formed so that the operator (5) of the agricultural working machine (25, 26) can discard the preselected driving route (14) and/or driving path (25, 26) and select any other driving path (25, 26, 80).

8. (withdrawn) The route planning system for agricultural working machines as defined in claim 7, wherein said means is formed so that when the operator (5) of the agricultural working machine (4) discards the preselected driving route (14) and/or driving path (25, 26), a new driving route (14') is determined, composed of driving paths (25, 26).

9. (currently amended) The route planning system for agricultural working machines <u>as defined in claim 1</u>, <u>wherein said</u> ~~comprising~~ means for assigning <u>formulate</u> a ~~defined working width to the agricultural working machine to generate driving paths for working a territory, and formulating~~ a working strategy <u>for said agricultural working machine</u>.

10. (previously presented) The route planning system for agricultural working machines as defined in claim 9, wherein said means is formed so that the working strategy includes connecting parallel driving paths (25, 26) and turning curves (37); incorporating the number and position

of additional agricultural working machines (3, 4) used on the territory (21) to be worked; consideration for the machine kinematics (12), the geometry of the territory (21) to be worked, consideration for harvested crop conditions (13); consideration for customer requests and implementing specified working sequences.

11. (withdrawn) The route planning system for agricultural working machines as defined in claim 9, wherein said means is formed so that it stores driving routes (14, 14') and working strategies for a territory (21) to be worked and recognizes these stored driving routes (14, 14') and working strategies (14, 14') when they are worked again and automatically accesses these stored driving routes (14, 14') and working strategies.

12. (withdrawn) The route planning system for agricultural working machines as defined in claim 9, wherein said means is formed so that the driving route (14, 14') generated using the driving paths (25, 26) is based on a master line (46), whereby the driving paths (25, 26) based on the master line (46) are offset from the master line (46) and from each other by nearly the working width (AB) of the agricultural working machine (3, 4) or a multiple thereof.

13. (withdrawn) The route planning system for agricultural working machines as defined in claim 11, wherein said means is formed so that the master line (46) can be drawn straight or curved and whereby each master line (46) is formed based on two path points (C, D) separated by a distance, and a virtual extension (47) of the path of the master line (46) extending through these path points (C, D) serves as a guide line (48).

14. (withdrawn) The route planning system for agricultural working machines as defined in claim 11, wherein said means is formed so that the master line (46) is defined by the operator (5) of the agricultural working machine (3, 4).

15. (withdrawn) The route planning system for agricultural working machines as defined in claim 13, wherein said means is formed so that the guide line (48) is used to automatically guide the agricultural working machine (3, 4).

16. (withdrawn) The route planning system for agricultural working machines as defined in claim 11, wherein said means is formed so that a number of generated path points on curved master lines (46) is reduced by running a computation algorithm (49).

17. (previously presented) The route planning system for agricultural working machines as defined in claims 9, wherein said means is formed

so that the generated driving paths (25, 26) first follow the shape of outer contours (23) of the territory (21) to be worked and subsequently extend nearly parallel to each other.

18. (previously presented) The route planning system for agricultural working machines as defined in claim 9, wherein said means is formed so that a length of the driving paths (25, 26) is determined by the outer contour (23) of the territory (21) to be worked.

19. (withdrawn) The route planning system for agricultural working machines as defined in claim 9, wherein said means is formed so that the driving paths (25, 26) on the driving route (14, 14') are extended virtually so far that, on the turn-around route (36), the agricultural working machine (3, 4) moved past the driving path (25, 26) on the territory (21) to be worked is moved so far away from the territory (21) that the agricultural working machine (25, 26) can be turned around without contacting non-worked ground (50).

20. (withdrawn) The route planning system for agricultural working machines as defined in claim 9, wherein said means is formed so that a transition at an end of one driving path (25, 26) to the next driving path (25, 26) is determined by a turn-around procedure (36) defined by a turn-around curve (37) that can be calculated.

21. (withdrawn) The route planning system for agricultural working machines as defined in claim 9, wherein said means is formed so that further driving paths (25, 26) are displayed to the operator (5) of the agricultural working machine (3, 4), at least at ends of the particular driving path (25, 26), and the operator (5) can select the next driving path (25, 26) to be worked, and the route planning system (1) automatically determines turn-around curve (37) for this turn-around route (36), and the turn-around curve can be calculated based on a large number of driving routes (25, 26).

22. (withdrawn) The route planning system for agricultural working machines as defined in claim 20, wherein said means is formed so that the operator (5) can select the driving path (25, 26) by operating a touch-screen monitor (51).

23. (withdrawn) The route planning system for agricultural working machines as defined in claim 9, wherein said means is formed so that the operator (5) of the agricultural working machine (3, 4) can shift the generated driving paths (25, 26, 37).

24. (currently amended) A route planning method for agricultural working machines, comprising the steps of assigning a defined working width to the agricultural working machines to generate driving routes in a territory; and carrying out dynamic adaptation of the planned driving route, <u>including automatically generating a new driving path for an agricultural working machine in response to an operator changing the machine's working sequence for the territory at any time, wherein the new driving path is automatically worked by the working machine</u> thereby ensuring that the driving route to be covered is flexibly adaptable to changing external conditions <u>but for</u> ~~such as~~ driving around obstacles <u>in working machine paths</u>, thereby largely relieving the operator of the agricultural working machine of the task of performing laborious steering maneuvers.

25. (withdrawn) The route planning method for agricultural working machines as defined in claim 24, wherein said carrying out includes dynamically adapting the planned driving route (14, 14') as a function of the actual machine position (31) and the actual machine orientation (33).

26. (withdrawn) The route planning method for agricultural working machines as defined in claim 24, wherein said carrying out includes providing the dynamic adaptation of the driving route (14, 14') permanently.

27. (withdrawn) The route planning method for agricultural working machines as defined in claim 24, further comprising generating the driving route (14, 14') based on a large number of driving paths (25, 26), and the driving paths (25, 26) are determined based on optimization criteria (11).

28. (withdrawn) The route planning method for agricultural working machines as defined in claim 24, further comprising selecting the next driving path (25, 26) to be worked based on optimization criteria (11).

29. (withdrawn) The route planning method for agricultural working machines as defined in claim 28, further comprising selecting the optimization criteria (11) "to be shortest driving route/processing time," "small proportion of unproductive auxiliary time," "short auxiliary drives between successive driving paths (25, 26) to be worked," "recognition and working of known driving routes (14, 14') and sequences," "short turn-around routes (36)," and "minimize routes between agricultural working machine (3, 4) and hauling vehicle (35)."

30. (withdrawn) The route planning method for agricultural working machines as defined in claim 24, further comprising discarding by the operator (5) of the agricultural working machine (25, 26) can the

preselected driving route (14) and/or driving path (25, 26) and select any other driving path (25, 26, 80).

31. (withdrawn) The route planning method for agricultural working machines as defined in claim 30, further comprising, when the operator (5) of the agricultural working machine (4) discards the preselected driving route (14) and/or driving path (25, 26), determining a new driving route (14') composed of driving paths (25, 26).

32. (withdrawn) The route planning method for agricultural working machines, comprising the steps of assigning a defined working width to the agricultural working machine to generate driving paths for working a territory, and formulating a working strategy.

33. (currently amended) A ~~The~~ route planning method for agricultural working machines, <u>comprising the steps of</u>

<u>assigning a defined working width to the agricultural working machine to generate driving paths for working a territory and formulating a strategy;</u>

<u>ensuring that the driving paths to be covered are flexibly adaptable to changing external conditions but for driving obstacles in the driving paths, including automatically generating a new driving path for an agricultural working machine in response to an operator changing the machine's working sequence for the territory at any time, wherein the new driving path is automatically worked by the working machine,</u>

~~as defined in Claim 32, further comprising~~ carrying out the working strategy to include connecting parallel driving paths (25, 26) and turning curves (37);

incorporating the number and position of additional agricultural working machines (3, 4) used on the territory (21) to be worked;

considering the machine kinematics (12), the geometry of the territory (21) to be worked;[[,]]

considering harvested crop conditions (13);

considering customer requests; and

implementing specified working sequences.

34. (withdrawn) The route planning method for agricultural working machines as defined in claim 32, further comprising storing driving routes (14, 14') and working strategies for a territory (21) to be worked and

recognizing these stored driving routes (14, 14') and working strategies (14, 14') when they are worked again and automatically accessing these stored driving routes (14, 14') and working strategies.

35. (withdrawn) The route planning method for agricultural working machines as defined in claim 32, further comprising generating the driving paths (25, 26) based on a master line (46), whereby the driving paths (25, 26) based on the master line (46) are offset from the master line (46) and from each other by nearly the working width (AB) of the agricultural working machine (3, 4) or a multiple thereof.

36. (withdrawn) The route planning method for agricultural working machines as defined in claim 34, further comprising drawing the master line (46) straight or curved and thereby forming each master line (46) based on two path points (C, D) separated by a distance, and providing a virtual extension (47) of the path of the master line (46) extending through these path points (C, D) to serve as a guide line (48).

37. (withdrawn) The route planning method for agricultural working machines as defined in claim 34, further comprising defining the master line (46) by the operator (5) of the agricultural working machine (3, 4).

38. (withdrawn) The route planning method for agricultural working machines as defined in claim 36, further comprising using the guide line (48) to automatically guide the agricultural working machine (3, 4).

39. (withdrawn) The route planning method for agricultural working machines as defined in claim 34, further comprising reducing a number of generated path points on curved master lines (46) by running a computation algorithm (49).

40. (currently amended) The route planning method for agricultural working machines as defined in claim 33, Claims 32, further comprising generating the driving paths (25, 26) so that they follow the shape of outer contours (23) of the territory (21) to be worked and subsequently extend nearly parallel to each other.

41. (currently amended) The route planning method for agricultural working machines as defined in claim 33, claim 32, further comprising determining a length of the driving paths (25, 26) by the outer contour (23) of the territory (21) to be worked.

42. (withdrawn) The route planning method for agricultural working machines as defined in claim 32, further comprising extending the driving paths (25, 26) on the driving route (14, 14') virtually so far that,

on the turn-around route (36), the agricultural working machine (3, 4) moved past the driving path (25, 26) on the territory (21) to be worked is moved so far away from the territory (21) that the agricultural working machine (25, 26) can be turned around without contacting non-worked ground (50).

43. (withdrawn) The route planning method for agricultural working machines as defined in claim 32, further comprising determining a transition at an end of one driving path (25, 26) to the next driving path (25, 26) by a turn-around procedure (36) defined by a turn-around curve (37) that can be calculated.

44. (withdrawn) The route planning method for agricultural working machines as defined in claim 32, further comprising displaying further driving paths (25, 26) to the operator (5) of the agricultural working machine (3, 4), at least at ends of the particular driving path (25, 26), so that the operator (5) can select the next driving path (25, 26) to be worked, automatically determining turn-around curve (37) for this turn-around route (36), and calculating the turn-around curve based on a large number of driving routes (25, 26).

45. (withdrawn) The route planning method for agricultural working machines as defined in claim 43, further comprising selecting by the operator (5) the driving path (25, 26) by operating a touch-screen monitor (51).

46. (withdrawn) The route planning method for agricultural working machines as defined in claim 32, further comprising shifting by the operator (5) of the agricultural working machine (3, 4) the generated driving paths (25, 26, 37).

REMARKS

[DESCRIBE THE CLAIMS AND GROUNDS FOR REJECTION AS INDICATED IN THE OFFICE ACTION. THE BOILERPLATE LANGUAGE IS FOUND IN BOLD.]

The present Amendment is submitted in response to the non-final Office Action dated [INSERT DATE].

The non-final Office Action states that after restriction/election, claims 1, 7, 10, 17, 18, 24, 33, 40, and 41 (directed to species B; Fig. 4) are pending for prosecution, that each of claims 1, 7, 10, 17, 18, 24,

33, 40, and 41 are provisionally rejected under 35 U.S.C. Section 101 and that each of claims 1, 10, 17, 18, 24, 33, 40, and 41 are rejected under 35 U.S.C. Section 102(b) as anticipated by [INSERT U.S. PATENT NUMBER] to [INSERT REFERENCE].

Applicants' representative called [EXAMINER NAME] on September 1, 2009, and discussed the restriction/election in view of the fact that claims 10, 17, and 18 depend from withdrawn claim 9, and claims 33, 40, and 41 depend from withdrawn claim 32.

[EXAMINER NAME] tentatively agreed that claim 9 should be amended to depend from pending independent claim 1, and therefore added to the claims elected for prosecution.

[EXAMINER NAME] further tentatively agreed that in view of the fact that independent claim 32 is quite broad, its subject matter should be added to claim 33, claim 32 should be cancelled without prejudice, and the dependencies of claims 40 and 41 should be changed to depend from claim 33 as amended.

Hence, applicants hereby amend claim 9 (understood to be no longer withdrawn) to depend from claim 1, amend claim 33 to independent form by including the subject matter of withdrawn claim 32, and amend claims 40 and 41 to depend from claim 33. Applicants have not cancelled withdrawn claim 32, as initially discussed.

Independent claims 1, 24, and 33 and dependent claims 7, 9, 10, 17, 18, 40, and 41 are pending for prosecution hereinafter.

[THE CLAIM AMENDMENTS OBVIATE THE PROVISIONAL DOUBLE PATENTING REJECTION.]

In response to the provisional rejection of claims 1, 7, 10, 17, 18, 24, 33, 40, and 41 under Section 101 over [SERIAL NUMBER], applicants have amended independent claims 1, 24, and 33 to include a limitation that the driving route to be covered by working machines is flexibly adaptable to changing external conditions but for driving obstacles in working machine paths.

In view of the fact that each of claims 1, 7, 10, 17, 18, 24, 33, 40, and 41 of [SERIAL NUMBER] include the limitation that the driving route to be covered by working machines is flexibly adaptable to changing external conditions such as driving around obstacles, applicants respectfully assert that amending all claims pending for prosecution in this application to include the negative limitation that the driving route to be covered by working machines is flexibly adaptable to changing external conditions but for driving obstacles in working machine paths **should obviate the provisional double patenting rejection under Section 101.**

Applicants, therefore, in view of the amendment to claims 1, 24, and 33, respectfully request withdrawal of the rejection of claims 1, 7, 10, 17, 18, 24, 33, 40, and 41 as provisionally rejected under 35 U.S.C. Section 101.

[THE CLAIMS ARE ANTICIPATED BY THE REFERENCE.]

In response to the rejection in view of [INSERT REFERENCE] under Section 102(b), applicants have further amended independent claims 1, 24, and 33 to include that the system automatically generates a new driving path for an agricultural working machine in response to an operator changing the machine's working sequence for the territory at any time, where the new driving path is automatically worked by the working machine.

Support for this limitation is found in applicants' Specification at page 6, lines 5–9.

[ANTICIPATION OCCURS WHEN EACH AND EVERY ELEMENT IN THE CLAIM IS FOUND IN THE REFERENCE.]

In view of the fact that amended independent claims 1, 24, and 33 recite at least this limitation, which [INSERT REFERENCE] does not, [INSERT REFERENCE] does not anticipate the invention as claimed. As such, [INSERT REFERENCE] is not a proper reference under 35 U.S.C. Section 102 pursuant to the guidelines set forth in the last paragraph of MPEP Section 2131, where it is stated that "a claim is anticipated only if each and every element as set forth in the claims is found, either expressly or inherently described, in a single prior art reference," and that "the identical invention must be shown in as complete detail as is contained in the . . . claim."

Amended independent claims 1, 24, and 33 are therefore patentable under 35 U.S.C. Section 102(b) over [INSERT REFERENCE]. Claims 7, 9, 10, 17, and 18 depend from independent claim 1 and are patentable therewith; claims 40 and 41 depend from independent claim 33 and patentable therewith. Applicants, therefore, respectfully

request withdrawal of the rejection of claims 1, 7, 10, 17, 18, 24, 33, 40, and 41 under 35 U.S.C. Section 102(b) over [INSERT REFERENCE].

Accordingly, the application as amended is believed to be in condition for allowance. Action to this end is courteously solicited. However, should the Examiner have any further comments or suggestions, the undersigned would very much welcome a telephone call in order to discuss appropriate claim language that will place the application in condition for allowance.

UNITED STATES PATENT AND TRADEMARK OFFICE

AMENDMENT

[INSERT DATE]

Commissioner for Patents

P O Box 1450

Alexandria, VA 22313-1450

Sir/Madam:

In response to the Office Action mailed on [INSERT DATE], applicants present the following amendments and remarks.

IN THE SPECIFICATION

Please amend the Title of the invention as follows:

[INSERT TITLE]

IN THE SPECIFICATION

Please insert the following Abstract of the Disclosure after page 11:

<u>Abstract of the Disclosure</u>

A public address system includes at least one monitoring device, a transmission medium and at least one level monitoring device. The monitoring device is connected to the transmission medium and powered by an AC voltage with a supply frequency present on the transmission medium. The monitoring device receives or transmits or transmits and receives a communication signal utilizing a carrier frequency upon the transmission medium. The carrier frequency is different from the supply frequency and the level monitoring device is adapted to monitor the level of the AC voltage.

IN THE SPECIFICATION

Please modify the Specification at page 2, the paragraph 5 beginning at line 27, as follows:

Further it is advantageous that the monitoring device comprises at least

one ~~on~~ power storage unit, especially at least one capacitor and/or at least one

battery and/or at least one rechargeable battery, because in this case the power consumption of the monitoring device is distributed evenly.

Please modify the Specification at page 3, line 20, as follows:

Brief Description of the Drawings ~~Drawing~~

Please modify the Specification at page 4, the paragraph 5 beginning at line 4, as follows:

FIG. 1 shows a block diagram of the configuration of a public address system, comprising loudspeakers 10, monitoring devices (M) 12, a transmission medium 18 and a control device (CD) 20. The control device 20 is integrated into an amplifier device (AD) 28. The amplifier device 28 further comprises an audio amplifier 22 and a network device 24 (ND). Communication links 26 connect the network device 24 to the audio amplifier 22 and the control device 20. The control device 20 and the output of the audio amplifier 22 are connected to the transmission medium 18. In the preferred embodiment the transmission medium 18 is a wire or a cable. The control device 20 controls all communication within the transmission medium 18 and also interfaces to the amplifier device 28 via an I2C interface. In the preferred embodiment the control device 20 is built in the amplifier device 28. The control device 20 is the master of the communication protocol. Every device, especially the monitoring devices (M) 12, 14 and/or the end of line monitoring devices (LD) 16, has its own unique address that is selected by the installer. The control device 20 polls all devices 12, 14, 16 to check if errors have occurred. Polling is an automatic, sequential testing of each connected device 12, 14, 16 in order to check its operational status. If a monitoring device 12, 14, and/or end of line monitoring device 16 responds, the device can report an error, especially a loudspeaker coil error and/or a power supply error and/or communication error and/or microprocessor error. If no error has occurred the monitoring device 12, 14, and/or end of line monitoring device 16 reports all is ok. If the monitoring device 12, 14, and/or end of line monitoring device 16 does not respond at all, it is assumed by the control device 20 that the device is malfunctioning or the transmission medium 18 is out of order. If an error occurs the control device 20 reports an error protocol to the network device 24. Each to be monitored loudspeaker 10 is associated with one monitoring device 12. In the preferred embodiment the public address system further comprises loudspeakers 12 not associated with a monitoring device. The function of the monitoring device 12 is to check if the loudspeaker coil is not an open connection. If an open connection is detected, the monitoring device 12 replies an error message to the control device 20. In the preferred embodiment the monitoring device 12 is powered by a 20 kHz pilot tone generated by the audio amplifier 22. The loudspeakers 10 are connected to the transmission medium 18 and the loudspeakers 10 receive audio frequencies within the audible frequency range, especially within the frequency range of 50 Hz to 18 kHz, from the audio amplifier 22. End of line monitoring of the transmission medium 18 is accomplished ether by a monitoring device 14 associated with a loudspeaker 10 and/or by an end of line monitoring device 16. The function of the monitoring device 14 and/or the end of line monitoring device 16 is to check if the transmission medium 18, e.g., the loudspeaker cable, is sill intact. If communication to the monitoring device 14 and/or the end of line monitoring device 16 fails, it is assumed by the control device 20, that the transmission medium 18 is out of order, e.g., the loudspeaker cable is broken or short-circuited. Therefore an end of line error is generated. In the preferred embodiment the end of line monitoring device 16 is powered by a 20 kHz pilot tone (AC voltage with a supply frequency preferably at 20 kHz) generated by the audio amplifier 22. In the preferred embodiment the monitoring devices 12, 14, 16 are polled a number of times, preferably between 5 and 10 times, to give every monitoring device 12, 14, 16 the possibility to response before it is assumed that a monitoring device 12, 14, 16 is not responding and conclude that the speaker line is out of order and/or a loudspeaker is disconnected and/or a loudspeaker coil is an open connection. This prevents reporting faulty errors to the network device 24.

Please modify the Specification at page 5, the paragraph 5 beginning at line 15, as follows:

FIG. 2 shows a block diagram of the monitoring device 12 and the control device 20. The monitoring device 12 is associated with a loudspeaker 10, whereas the control device 20 is associated with the audio amplifier 22 and the network device 24. The loudspeaker 10, the monitoring device

12, the control device 20 and the output of the audio amplifier 22 are connected to the transmission medium 18, wherein the transmission medium 18 in the preferred embodiment is a loudspeaker wire. The control device 20 comprises a microprocessor (μ) 30, a transmitter (TX) 32, a receiver (RX) 34, a filter of the transmitter (TX Filter) 36, a filter of the receiver (RX Filter) 38 and a level monitoring device (LM) 40. On the other side the monitoring device 12 comprises a loudspeaker monitoring device (LSPK M) 42, a microprocessor (μ) 44, a transmitter (TX) 46, a receiver (RX) 48, a power supply (PS) 50, a filter of the transmitter (TX Filter) 52, a filter of the receiver (RX Filter) 54 and a filter of the power supply (PS) 56. The microprocessor 44 of the monitoring device 12 controls the communication and supervises the loudspeaker 10. The power supply 50 of the monitoring device 12 or a power supply of the end of line monitoring device have got the two main functions. The first function is to convert the voltage from the power supply filter 56 to voltage levels used on the monitoring device 12, whereas the second function is to store energy during reception and deliver extra energy during transmission, because when the monitoring device 12 is not transmitting, the power consumption is low, whereas the power consumption increases during sending. To prevent varying load, a power storage unit is used to distribute the power consumption evenly. In the preferred embodiment the power storage unit is a capacitor. In the preferred embodiment this is achieved by delivering a constant power from the power supply filter 56. When the device is not transmitting, surplus power is stored in the capacitor, but when the device starts transmitting this stored power is used. Because the power supply is derived from a 20 kHz pilot tone, a filter of the power supply 56 takes care that all other signals, especially the audio signal and the communication signal, are not unnecessary loaded. The transmitter 46 of the monitoring device 12 encodes and modulates the data. A filter of the transmitter 52 processes the encoded and modulated data. The filter of the transmitter 52 comprises an amplifier, a transformer, and a band-pass filter in order to superimpose the 75 kHz communication signal on the normal audio signals present on the transmission medium 18. The transmitter 46 is disconnected from the transmission medium 18 while no transmission takes place. Therefore only one transmitter 46 of all monitoring devices 12 is connected to the transmission medium 18 at the same time. The filter of the receiver 54 is a band-pass filter allowing 75 kHz communication signals passing the filter while other unwanted signals are attenuated. In order to prevent influencing the communication signal and/or the audio signal and/or the pilot tone of the power supply the input impedance of the filter of the receiver 54 is high. The receiver 48 itself demodulates and decodes the data and makes the data available to the microprocessor 44. Further the loudspeaker monitoring device 42 of the monitoring device 12 supervises the current through the coil of the loudspeaker 10. Because of the presence of the 20 kHz pilot tone a minimum current always flows through the coil of the loudspeaker 10. A too low level of this current is based on a fault of the loudspeaker 10 and this loudspeaker status is reported to the microprocessor 44 of the monitoring device 12. The microprocessor 30 of the control device 20 controls the communication, supervises the power supply, and communicates with the network device 24. The control device 20 receives its power supply directly from the network device 24 and/or the amplifier device. The configuration and the function of the transmitter 32, the receiver 34, the filter of the transmitter 36, and the filter of the receiver 38 is almost the same as the transmitter 46, the receiver 48, the filter of the transmitter 52, and the filter of the receiver 54 of the monitoring device 12. Further the level monitoring device 40 measures the actual level of the AC power voltage, in the preferred embodiment the level of the 20 kHz pilot tone, and reports the measurements to the microprocessor 30. This level is depending on the load of the transmission medium 18 and varies in different installations and situations and is depending on used cable (wire) cable length, used type of loudspeakers 12, and the number of loudspeakers 12. In the preferred embodiment the microprocessor 30 of the control device 20 and/or the microprocessor 44 of the monitoring device are low power microprocessors with sleep mode and they are configured to sleep during no communication.

IN THE CLAIMS

This **Listing of Claims** will replace all prior versions, and listings, of claims in the application:

Listing of Claims:

1. (currently amended) Public address system, comprising

at least one monitoring device, ~~and~~

a transmission medium, and

at least one level monitoring device;

wherein the monitoring device is connected to the transmission medium, and ~~wherein the monitoring device~~ is powered by an ~~a~~ AC voltage with a supply frequency present on the transmission medium,

wherein the monitoring device receives or transmits or ~~and/or~~ transmits and receives a communication signal utilizing a carrier frequency upon the transmission medium, wherein the carrier frequency is different from the supply frequency, <u>and</u>

<u>wherein the level monitoring device is adapted to monitor the level of the AC voltage</u>.

2. (original) Public address system according to claim 1, characterized in that the public address system further comprises at least one loud-speaker, wherein the loudspeaker receives audio signals present on the same transmission medium.

3. (previously presented) Public address system according to claim 1, characterized in that the monitoring device is connected to at least one end of the transmission medium.

4. (cancelled)

5. (previously presented) Public address system according to claim 1, characterized in that at least one loudspeaker line is the transmission medium.

6. (currently amended) Control device of a public address system, pref-erably of the public address system according to claim 1, characterized in that the control device is adapted to poll the monitoring devices by receiv-ing <u>or receiving or</u> ~~and/or~~ transmitting <u>and receiving</u> the communication signal utilizing the carrier frequency upon the transmission medium.

7. (currently amended) Control device according to claim 6, character-ized in that the control device is adapted to poll the monitoring devices a number of <u>times</u> ~~time~~ before the control device reports an error.

8. (currently amended) Monitoring device of a public address system, ~~preferably of the public address system~~ according to <u>claim 1, further</u> ~~one of the preceding claims~~, comprising

at least one power unit, wherein the power unit is adapted to power the monitoring device by a AC voltage with a supply frequency present on a transmission medium,

at least one receiver and/or at least one transmitter, wherein the receiver <u>or the transmitter or the</u> ~~and/or~~ the transmitter <u>and the receiver are</u> ~~is~~ adapted to receive <u>or transmit or</u> ~~and/or~~ transmit <u>and receive</u> a communication signal utilizing a carrier frequency upon the same trans-mission medium, wherein the carrier frequency is different from the sup-ply frequency, <u>and</u>

a microprocessor that operates in an operational mode and in a sleep mode, wherein the microprocessor is programmed to operate in the sleep mode during a time in which there is no communication.

9. (currently amended) Monitoring device according to claim 8, characterized in that the transmitter transmits status information of the monitoring device or the loudspeaker or the monitoring device and and/or the loudspeaker.

10. (currently amended) Monitoring device according to claim 8, characterized in that the transmitter is being activated upon transmission.

11. (currently amended) Monitoring device according to claim 1, further comprising characterised by at least one power storage unit, wherein the power storage unit is preferably at least one capacitor or at least one battery or at least one capacitor and and/or at least one battery, wherein the at least one battery may comprise and/or at least one rechargeable battery.

12. (previously presented) Monitoring device according to claim 1, characterized in that the monitoring device is associated with at least one loudspeaker, wherein the monitoring device monitors the functioning of the loudspeaker, especially by supervising a current through at least one coil of the loudspeaker.

13. (currently amended) Monitoring device according to claim 1, characterized in that the supply frequency is at least one of the group consisting of: above 18 kHz, below 100 Hz, above 18 kHz and and/or below 100 Hz, above 18 kHz and and/or direct current (DC), and below 100 Hz and direct current (DC) especially at 20 kHz.

14. (previously presented) Monitoring device according to claim 1, characterized in that the carrier frequency is above the audible audio band, especially at 75 kHz.

15. (previously presented) Monitoring device according to claim 1, characterized in that the communication signal is a phase shift keying (PSK) modulated signal, especially a differential phase shift keying (DPSK) modulated signal or a binary differential phase shift keying (binary DPSK) modulated signal.

16. (previously presented) Monitoring device according to claim 1, characterized in that the communication signal contains biphase coded data.

17. (new) Monitoring device according to claim 1, characterized in that the supply frequency includes 20 kHz and direct current (DC).

REMARKS

[DESCRIBE THE CLAIMS AND GROUNDS FOR REJECTION AS INDI-CATED IN THE OFFICE ACTION. BOILERPLATE LANGUAGE IS INDI-CATD IN BOLD; MAKE SURE THAT IT APPLIES TO YOUR CASE.]

The present Amendment is submitted in response to the Office Action mailed in this case on [INSERT DATE].

[THE EXAMINER OBJECTS TO THE DRAWINGS, ABSTRACT, AND TITLE.]

The Office Action objects to the drawings, objects to the Specification with respect to the Abstract and the Title, objects to the disclosure for a number of formal errors, rejects claims 1–16 under 35 U.S.C. Section 112, second paragraph, rejects claims 1–3 and 5–14 under Section 102(b) over [INSERT PUBLICATION NUMBER] to [INSERT REFERENCE] and rejects claims 4, 15, and 16 under Section 103(a) over [INSERT REFERENCE].

In response to the objection to the drawings, applicants amend Figs. 1 and 2, substantially in accordance with the Examiner's suggestions. The Amendments to Figs. 1 and 2 are shown in the REPLACEMENT SHEET accompanying this Amendment. For that matter, applicant also amends the Specification as shown above to reflect the amendments to Figs. 1 and 2. In view of the amend-ments to Figs. 1 and 2 (as shown in the REPLACEMENT SHEET), and the amendments to the Specification (as shown above), appli-cants respectfully asserts that the drawings are in proper form and, respectfully request withdrawal of the objection to the drawings therefore.

In response to the objection with respect to the Abstract, appli-cants present the Abstract of the Disclosure (as shown above) and respectfully request withdrawal of the objection therefore.

In response to the objection for the formal errors, applicants amend the Specification (as shown above), substantially in accor-dance with the Examiner's suggestions and, respectfully request withdrawal of the objection therefore.

In response to the objection to the Title, applicants amend the Title (as shown above) and, respectfully request withdrawal of the objection therefore.

In response to the rejection of claims 1–16 under Section 112, second paragraph, applicants amend claims 1, 6–11, and 13, and present new claim 17, as shown above, substantially in accordance with the Examiner's comments. In view of the amendments, claims 1–16 are believed to be patentable under Section 112, second paragraph, and applicants respectfully request withdrawal of the rejection.

[THE EXAMINER INDICATES THAT THE CLAIMS ARE ANTICIPATED BY THE REFERENCE.]

Turning now to the art rejections, [INSERT REFERENCE] discloses a speaker system (100; Fig. 1) with multiple speakers connected to a central amplifier speaker line that is controlled from a central location using a master/slave protocol by master control units 102A-D. At least one master control unit 102A is connected to a tone generator/mixer 122, a plurality of amplifiers 124, a bus 126, a bus 128, a plurality of remote units 130, and a plurality of speakers 152.

To support the rejection under Section 102(b), the Examiner asserts that [INSERT REFERENCE]'s public address system comprises a monitoring device (remote unit) 130, which is connected to a transmission medium. The transmission medium is defined as the lines connecting both computer control 102A and amplifiers 124 with the monitoring device (remote unit) 130.

Applicants see that computer control 102A is connected to monitoring device (remote unit) 130 by two electrical paths. The first electrical path is via bus 128, tone generator/mixer 122, amplifiers 124 to the monitoring device (remote unit) 130. The second electrical path is via audio lines 140 to the monitoring device (remote unit) 130. Par. [028] describes that the computer control 102A generates a signal that is provided over bus 128 to the 35 Hz tone generator/mixer 122. Upon receipt, the 35 Hz generator/mixer decodes the signal and generates an independent 35 Hz analog power signal to power the monitoring device (remote unit) 130 via the 35 Hz amplifiers 124.

While the Examiner asserts that the monitoring device (remote unit) 130 receives and transmits a communication signal utilizing a "radio frequency" carrier frequency upon the transmission medium, where the carrier frequency (100 kHz) is different from the supply frequency (35 Hz), applicants again respectfully disagree. Monitoring device (remote unit) 130 does receive signals from both the 35 Hz tone generator/mixer 122 and the computer control 102A, both signals appear to be 35 Hz. **Nowhere does [INSERT REFERENCE] suggest** that RF transceiver 144 as shown in Fig. 3 transmits an RF signal at 100 kHz.

In more detail, the computer control 102A sends a 35 Hz command signal by audio line connection 104 that is received in RF transformer 136. The RF transformer 136 communicates the signal to RF transceiver 144, wherein it is demodulated and communicated to 9600 baud modem 148. The RF transceiver operates on the same frequency as the signal it receives from the transformer, which is the same frequency generated by the computer control 102A. The digital signal from modem 148 is sent to microcontroller 148 to execute the command. **Again, applicants do not see that [INSERT REFERENCE] discloses** that the RF transceiver 144 transmits a 100 kHz signal to 9600 baud modem 148, or receives a 100 kHz signal therefrom.

While applicants believe the claim 1 as filed is distinguishable from [INSERT REFERENCE] under Section 102(b), they amend claim 1 to include the subject matter of claim 4 (now cancelled) in order to further prosecution.

Amended independent claim 1 now calls out a public address system including at least one monitoring device, a transmission medium, and at least one level monitoring device, wherein the monitoring device is connected to the transmission medium, is powered by an AC voltage with a supply frequency present on the transmission medium, and receives or transmits or transmits and receives a communication signal utilizing a carrier frequency upon the transmission medium, wherein the carrier frequency is different from the supply frequency and wherein the level monitoring device is adapted to monitor the level of the AC voltage.

For that matter, applicants further amend claim 8 to include a microprocessor that operates in an operational mode and in a sleep mode and is programmed to operate in the sleep mode during a time in which there is no communication.

[THE APPLICANT STATES THAT THE REFERENCE DOES NOT TEACH OR SUGGEST THE INVENTION.]

[INSERT REFERENCE] does not teach or suggest the use of level monitoring devices adapted to monitor the level of the AC voltage.

[THE EXAMINER STATES THAT IT WOULD HAVE BEEN OBVIOUS TO MODIFY THE REFERENCE TO ACHIEVE THE PRESENT INVENTION.]

In more detail, while in the rejection of claim 4 under Section 103(a) asserts that [INSERT REFERENCE] discloses at least one level monitoring device which, instead of monitoring the level of the AC voltage, monitors the current between the remote unit 130 and the speakers (par. [0033]), **and that it would have been obvious to modify [INSERT REFERENCE] to perform voltage monitoring, instead of current monitoring, applicants respectfully disagree.**

[INSERT REFERENCE] at par. [0045] discloses that the microcontroller 140 in the monitoring device (remote unit) 130 communicates the settings to the tap control and speaker fault sense circuit 142, which adjusts relays that change the transformer settings on speaker transformer 134. The tap control and speaker fault sense circuit 142 also monitors the current between the speaker 152 and the monitoring device (remote unit) 130 using a current transformer. If a drop in current is detected by the current transformer within the tap control and speaker fault sense circuit 142, the monitoring device (remote unit) 130 informs the master computer control station 102.

Such current monitoring is to determine whether individual speakers are being driven effectively by the transformer 134 together with the tap control and speaker fault sense device 142. Applicants monitor the level of the AC voltage powering the monitoring device, for example (without limitation) the pilot tone (20 kHz AC voltage).

[INSERT REFERENCE]'s use of the current transformer in the tap control and speaker fault sense circuit 142 to monitor the current between the speaker 152 and the monitoring device (remote unit) 130 is not equivalent to applicants claimed level monitoring device for monitoring AC voltage level. Not only are the different in that a current transformer is different that a voltage level monitoring device, they are not used to perform the same functions. **Put another way**, the current transformer does not monitor a current delivered to the monitoring device (remote unit) 130.

[THE APPLICANT STATES THAT ALTHOUGH IT WOULD HAVE BEEN POSSIBLE TO MODIFY THE REFERENCE, DOING SO WOULD SIGNIFICANTLY MODIFY THE REFERENCE'S INTENDED OPERATION.]

And while it would not be impossible to modify [INSERT REFERENCE] to monitor the current level of the monitoring device (remote unit) 130, or to even monitor a voltage level of signals driving the monitoring device (remote unit) 130, **as claimed, doing so would significantly modify [INSERT REFERENCE]'s intended operation (see** *In re Gordon,* **221 USPQ 1125 (Fed. Cir. 1984)), and possibly prevent [INSERT REFERENCE] from operating as intended.** That is, the master computer control station 102 is not configured to process a signal identifying a current or voltage level driving the monitoring device (remote unit) 130 (see *In re Ratti,* 123 USPQ 349 (CCPA 1959)). **Either case compels a legal conclusion that the proposed modifications cannot be obvious under the law; MPEP 2143.01.**

[THERE MUST BE SOME ARTICULATED REASONING WITH SOME RATIONAL UNDERPINNING IN ORDER TO SUPPORT A FINDING OF OBVIOUSNESS.]

Perhaps as importantly, however, the Examiner supplies no reason as to why it would have been obvious to substitute applicants' claimed AC voltage level monitoring device for [INSERT REFERENCE]'s current monitoring transformer in the tap control and speaker fault sense circuit 142, within monitoring device (remote unit) 130. **An obviousness rejection cannot be effected/sustained without some articulated reasoning with some rational underpinning to support the legal conclusion of obviousness;** *KSR v. Teleflex, Inc.* **82 USPQ2d. 1385, 1396 (2007).**

[AN OBVIOUSNESS REJECTION BASED ON "DESIGN CHOICE" MUST PRESENT A CONVINCING LINE OF REASONING AS TO WHY THE SKILLED ARTISAN WOULD HAVE FOUND THE MERE DESIGN CHOICE OBVIOUS.]

While the rejection implies that it would have been an obvious design choice to use AC voltage rather than current **to support that claim 4 is directed to obvious subject matter, an obviousness rejection based in "design choice" must present a convincing line of reasoning as to why the skilled artisan would have found the mere design choice obvious;** *Ex Parte Clapp,* **227 USPQ 972, 973 (BPAI 1985).** *Clapp* requires the Examiner to first review the Specification to ascertain what purpose the limitation (AC voltage level monitoring device) **serves (see at least page 2, lines 9–12), and then an explanation of the reasoning as to why [INSERT REFERENCE]'s** current transformer in the tap control and speaker fault sense circuit 142 performs AC voltage level monitoring **equally as well. The rejection does not appear to supply either.**

[THE APPLICANT STATES THAT THE CLAIMS ARE NON-OBVIOUS.]

Applicants respectfully assert, therefore, that it would not have been obvious to modify applicants' invention with [INSERT REFERENCE]'s current transformer in the tap control and speaker fault sense circuit 142, **and that amended independent claim 1, and claims 2, 3 and 5–14 that depend from claim 1 are patentable under both Section 102(b) and Section 103(a) over [INSERT REFERENCE]. Applicants respectfully request withdrawal of the rejections therefore.**

As claims 15 and 16 include at least the limitations included in amended independent claim 1, applicants further respectfully assert that claims 15 and 16 are patentable under Section 103(a) over [INSERT REFERENCE] for at least the same reasons and request withdrawal of the rejection of claims 15 and 16, therefore.

The application in its amended state is believed to be in condition for allowance. Action to this end is courteously solicited. Should the Examiner have any further comments or suggestions, the undersigned would very much welcome a telephone call in order to discuss appropriate claim language that will place the application into condition for allowance.

Opinions

Patentability Opinions

As its name suggests, a patentability opinion is a legal opinion that analyzes whether an invention is likely to satisfy the requirements for the grant of a new patent. In rendering a patentability opinion, the patent attorney will conduct a prior art search and review the relevant prior art to determine whether the invention appears to meet the patent office requirements of usefulness, novelty, and non-obviousness.

Freedom to Operate Opinions

A freedom to operate opinion is typically solicited when a business is planning to release a new product or process. The primary objective of a freedom to operate opinion is to determine whether there are any problematic patents in the field of technology prior to releasing the product or process to market. In other words, the business wants to know their risk of being sued for patent infringement if they launch a new product or process. When relevant in-force prior patents are located, the freedom to operate opinion is generated by considering the entire file history of the patent, not just the patent itself.

Infringement Opinions

An infringement opinion is somewhat similar to a freedom to operate opinion with one primary exception: The patent that creates the basis for liability is known and does not have to be located through a prior art search. Typical infringement opinions will construe the claims, and then compare the claims to the product to determine whether the product infringes any of the claims. As with a freedom to operate opinion, an infringement opinion must be made in light of the patent file history to

97

determine whether any statements made during prosecution could limit or expand the enforceability of the claims.

Invalidity Opinions

Patent validity opinions are useful when concerned about infringement or preparing for a patent infringement lawsuit. If the patent is invalid or unenforceable, then you may be entitled to practice the claimed invention without liability for infringement. Patent validity opinions are formed after performing an in-depth patent validity search. For a patent holder, you may want a validity opinion before undertaking potentially significant litigation costs enforcing your patent rights or before entering into licensing negotiations. Similar to a non-infringement opinion, obtaining an appropriate patent invalidity opinion demonstrates good faith and reasonableness and helps avoid charges of willful infringement and award of increased damages to patentee during defensive patent infringement actions.

ATTORNEY/CLIENT PRIVILEGED

<u>**VIA FEDERAL EXPRESS**</u>   <u>**ATTORNEY WORK PRODUCT**</u>

Re: Invalidity Study of [INSERT] patent

Dear Sir/Madam:

In response to your request, we have considered whether the claims of [INSERT] patent are invalid. You have asked us to review and analyze all of the claims of the [INSERT] patent with respect to invalidity. A copy of the [INSERT] patent is attached as Exhibit [INSERT].

As requested, we have analyzed the [INSERT] patent, including its claims, specification, and prosecution history. We have also analyzed various prior art references located by our own search of the prior art. These documents are provided respectively in Appendices [INSERT].

A brief summary of our opinion follows, after which the reasons for our conclusions are set forth in detail.

SUMMARY OF OPINION

Based upon our careful study of the [INSERT] patent, its prosecution history and the prior art, it is our opinion that all of the claims of the [INSERT] patent are invalid in view of the prior art and the knowledge of a person of ordinary skill in the art. The [INSERT] patent contains [INSERT NUMBER OF CLAIMS], i.e., claims 1-[INSERT], of which, claims 1 and [INSERT] are drafted in independent form. Claims [INSERT] are invalid for at least the following reasons:

It is our opinion that claims [INSERT] are invalid under 35 U.S.C. Section 103(a) as obvious in view of [INSERT REFERENCES].

It is also our opinion that claims [INSERT] are not infringed by the relevant product.

The [INSERT REFERENCES] qualify as prior art under 35 U.S.C. Section 103(a) because they were all published more than one year prior to the earliest effective filing date of the [INSERT] patent—namely, [INSERT FILING DATE].

Further, the [INSERT REFERENCES] were not of record during prosecution of the [INSERT] patent, and thus, were not considered by the Examiner.

To the extent that a well-informed court might find a description of any claim element missing from these prior art references, such element would have been known to a person of ordinary skill in the art at the relevant time and likely would have been an obvious variation to the systems and methods disclosed in those references.

We set out the grounds for our opinions in detail as follows.

I. The [INSERT PATENT]

A. Background

[INSERT] patent entitled [INSERT TITLE], is assigned to [INSERT ASSIGNEE].

[INSERT ISSUE DATE AND DETAILS OF THE 123 PATENT]

B. The Relevant Teachings of the [INSERT PATENT]

[INSERT TECHNICAL DISCUSSION OF PATENT]

II. THE FILE HISTORY OF THE [INSERT] PATENT

We will briefly summarize the patent file history of the [INSERT] patent, the reexamination of the [INSERT] patent, and the successor patents of the [INSERT] patent.

A. Summary of the [INSERT] patent File History

The [INSERT] patent issued on [INSERT ISSUE DATE] from U.S. Patent Application Serial No. [INSERT NUMBER] (the [INSERT] patent application), filed on [INSERT FILING DATE]. [INSERT NAMES OF INVENTORS] are the named inventors of the [INSERT] patent. As originally filed, the [INSERT] patent application contained independent claims [INSERT] and was entitled [INSERT].

On [INSERT], Applicants filed "Preliminary Amendment A" in which new claims [INSERT] were added, of which claims [INSERT] were drafted in independent form.

[DISCUSS RELEVANT PORTIONS OF FILE HISTORY]

III. THE PATENT LITIGATIONS INVOLVING THE ORIGINAL [INSERT] PATENT, THE REEXAMINED [INSERT] AND THE [INSERT] PATENT

A. [INSERT] v. [INSERT]

On [INSERT DATE], [INSERT PARTY] filed a complaint against [INSERT PARTY] in the [INSERT COURT] for alleged infringement of the [INSERT] patent.

[INSERT DISCUSSION, IF ANY, OF ANY PENDING AND PAST LAWSUITS CONCERNING THE PATENT]

IV. THE PRIOR ART

A. [INSERT—NAME OF PRIOR ART]

U.S. Patent No. [INSERT] is entitled "[INSERT]" and issued on [INSERT] based on U.S. Patent Application Serial No. [INSERT], which was filed on [INSERT]. The [INSERT] patent names as inventors [INSERT NAMES OF INVENTORS] and is assigned to [INSERT]. [INSERT REFERENCE] was of record during reexamination of the [INSERT] patent but was never applied by the Examiner. A copy of the patent is attached as Appendix [INSERT]. [INSERT] qualifies as prior art of the [INSERT] patent under 35 U.S.C. Section 102(e). First, [INSERT REFERENCE] filing date predates the earliest effective filing date of the [INSERT] patent.

[INSERT] is directed to [INSERT – DISCUSSION OF PRIOR ART]

B. [INSERT—PRIOR ART REFERENCE]

[INSERT] entitled "[INSERT]" was published on [INSERT DATE] as Publication No. [INSERT] (hereinafter "[INSERT]"). A copy of the publication is attached as Appendix [INSERT]. [INSERT] qualifies as prior art under 35 U.S.C. Section 102(b) (i.e., it was published on [INSERT], which is more than one year before the earliest effective filing date of the [INSERT] patent—namely, [INSERT]). [INSERT] was not of record during prosecution of the application leading to the [INSERT] patent.

[INSERT] describes a method and system for [INSERT—DISCUSSION OF PRIOR ART]

V. APPLICABLE LEGAL PRINCIPLES

A. Applicable Law of Patent Validity

[THE BURDEN OF PROVING INVALIDITY RESTS ON THE PARTY CHALLENGING THE PATENT. INVALIDITY MUST BE PROVED BY CLEAR AND CONVINCING EVIDENCE.]

Under 35 U.S.C. Section 282 of the U.S. Patent Laws, a patent is presumed to be valid. The burden of proving invalidity rests on the party challenging the patent. Invalidity must be proved by clear and convincing evidence. *Microsoft Corp. v. i4i L.P.*, 131 S.Ct. 2238, 2252 (2011). When a challenger relies on prior art that was not considered by the PTO, this burden remains the same. *Id*. The issue of whether a prior art reference was considered by the PTO is a factor to consider when determining whether the burden is met. *Id*. at 13 and note. 10. The invalidity of an issued patent may be proven using prior art references the Examiner had considered. *See Tyler Refrigeration v. Kysor Indus. Corp.*, 777 F.2d 687, 690 (Fed. Cir. 1985).

[AN INVENTION MUST BE NEW, NON-OBVIOUS AND USEFUL.]

To be valid, an invention claimed in a patent must satisfy at least the statutory requirements of being useful, new, and non-obvious. 35 U.S.C. Sections 101 (utility), 102 (novelty) and 103 (obviousness).

1. Claim Construction

[CLAIM CONSTRUCTION IS DECIDED BY THE COURT.]

Before the validity of a claim can be evaluated, its meaning and scope must be determined. This process is called claim construction, and is an issue of law that is decided by the court (i.e., the judge), even in the case of a jury trial. *Markman v. Westview Instruments, Inc.*, 517 U.S. 370, 372 (1996).

[CLAIM LANGUAGE IS INTERPRETED IN LIGHT OF THE CLAIMS, SPECIFICATION, AND PROSECUTION HISTORY.]

Claim language is interpreted in light of the intrinsic evidence of record, i.e., the claims, the specification, and the prosecution history. "Such intrinsic evidence is the most significant source of the legally operative meaning of disputed claim language." *Vitronics Corp. v. Conceptronic, Inc.*, 90 F.3d 1576, 1582 (Fed. Cir. 1996). Within this intrinsic evidence, "[t]he appropriate starting point . . . is always the language of the asserted claim itself." *Comark Communications, Inc. v. Harris Corp.*, 156 F.3d 1182, 1186 (Fed. Cir. 1998). That is, a claim term should be given its ordinary meaning unless (1) the patentee expressly defines the term in the specification or (2) the term has no clear meaning to those of ordinary skill in the art. *Johnson Worldwide Assocs. V. Zebco Corp.*, 175 F.3d 985, 989 (Fed. Cir. 1999); *see also Helmsderfer v. Bobrick Washroom Equipment, Inc.*, 527 F.3d 1379, 1381 (Fed. Cir. 2008). ("A patentee may act as its own lexicographer and assign to a term a unique definition that is different from its ordinary and customary meaning; however, a patentee must clearly express that intent in the written description."); *Irdeto Access, Inc. v. Echostar Satellite Corp.*, 383 F.3d 1295, 1300 (Fed. Cir. 2004). ("Moreover, if a disputed term has no previous meaning to those of ordinary skill in the prior art, its meaning, then, must be found elsewhere in the patent.") (citations omitted); *Multiform Desiccants, Inc. v. Medzam, Ltd.*, 133 F.3d 1473, 1477 (Fed. Cir. 1998) (inventors may bestow special meaning to a term in the specification), *J.T. Eaton & Co. v. Atlantic Paste & Glue Co.*, 106 F.3d 1563, 1568, 41 U.S.P.Q.2d 1641, 1646 (Fed. Cir. 1997) (if a claim term is without a meaning to those of ordinary skill in the art, its meaning "must be found elsewhere in the patent.") and *York Prods., Inc. v. Central Tractor Farm & Family Ctr.*, 99 F.3d 1568, 1572 (Fed. Cir. 1996) ("Without an express intent to impart a novel meaning to claim terms, an inventor's claim terms take on their ordinary meaning." One resource for determining the ordinary meaning of a claim term is a dictionary.)

[WE MUST REVIEW THE PROSECTION HISTORY TO SEE IF THERE ARE ANY TERMS DIFFERENT FROM THE ORDI-NARY MEANING.]

It is also necessary to review the prosecution history to determine whether the inventor has used any terms in a manner different from their ordinary meaning. See, e.g., *Verizon Services Corp. v. Vonage Holdings Corp.*, 503 F.3d 1295, 1307 (Fed. Cir. 2007) (claim term narrowed by arguments and amendments made during the prosecution); *Southwall Techs., Inc. v. Cardinal IG Co.*, 54 F.3d 1570, 1576 (Fed. Cir. 1995), *cert. denied*, 116 S.Ct. 515, 133 L.Ed.2d 424 (1995) (same). For example, a patentee can limit the interpretation of claims by excluding certain interpretations during prosecution in order to obtain claim allowance. *Alza Corp. v. Mylan Laboratories, Inc.*, 391 F.3d 1365, 1372 (Fed. Cir. 2004); *Standard Oil Co. v. American Cyanamid Co.*, 774 F.2d 448, 452 (Fed. Cir. 1985). The prosecution history can also provide evidence as to the meaning of the claims because "arguments made during prosecution regarding the meaning of a claim term are relevant to the interpretation of that term in every claim of the patent absent a clear indication to the contrary." *CVI/BETA Ventures, Inc. v. Tura LP*, 112 F.3d 1146, 1155, 42 U.S.P.Q.2d 1577, 1583 (Fed. Cir. 1997) (quoting *Southwall Techs. Corp. v. Cardinal IG Co.*, 54 F.3d 1570, 1579, 34 U.S.P.Q.2d 1673, 1679 (Fed. Cir.)), *cert. denied*, 133 L.Ed.2d 424, 116 S.Ct. 515 (1995).

[THE PROSECUTION HISTORY HELPS TO DETERMINE WHETHER THE PATENTEE IS CONTRADICTING HIS POSI-TION TAKEN DURING PROSECUTION.]

The prosecution history also prevents the patentee from interpreting the claims in a manner inconsistent with positions taken during prosecution. Prosecution history "limits the interpretation of claim terms so as to exclude any interpretation that was disclaimed during prosecution." Id. at 1155; *Microsoft Corp. v. Multi-Tech Sys., Inc.*, 357 F.3d 1340, 1349 (Fed. Cir. 2004) ("A patentee may also limit the scope of the claims by disclaiming a particular interpretation during prosecution.").

[CLAIM TERMS USUALLY CANNOT BE NARROWED BY REFERENCE TO THE WRITTEN DESCRIPTION.]

It is well-settled that "[a]n inventor must describe what he conceives to be the best mode, but he is not confined to that." *Continental Paper Bag Co. v. Eastern Paper Bag Co.*, 210 U.S. 405 (1908). Accordingly, "claim terms cannot be narrowed by reference to the written description or prosecution history unless the language of the claims invites reference to those sources." *Johnson Worldwide*, 175 F.3d at 989. Nonetheless, a claim term may be narrowly construed where the specification fails to support a broader interpretation. For example, in *Kemco Sales Inc. v. Control Papers Co.*, 208 F.3d 1352, 1363 (Fed. Cir. 2000) the Federal Circuit held that the "closing means" was properly construed to require that the "closing means" be secured to the outside of an envelope material because a "more expansive interpretation is simply not supported by any disclosure in written description."

[THE SPECIFICATION CAN BE USED TO DISCOVER THE INVENTOR'S INTENT.]

Alternately, "it is entirely proper to use the specification in order to determine what the inventor meant by terms and phrases in the claims." *Laitram Corp. v. Morehouse Indus., Inc.*, 143 F.3d 1456, 1462 (Fed. Cir. 1998). For example, remarks describing the claimed invention and the cited prior art made by an attorney during prosecution of a patent may properly be used to interpret claim language. *Desper Prods. Inc. v. QSound Labs Inc.*, 157 F.3d 1325, 1338-40 (Fed. Cir. 1998); see also *CVI/Beta Ventures*, 112 F.3d at 1158 (statements made during prosecution "may commit to particular meaning for patent term which is then binding").

[EXTRINSIC EVIDENCE CANNOT BE USED TO CONTRADICT THE MEANING OF THE CLAIM LANGUAGE.]

Although extrinsic evidence may be used to resolve any ambiguity in the claims, it cannot be used to contradict the established meaning of the claim language. *Gart v. Logitech, Inc.*,

254 F.3d 1334, 1340 (Fed. Cir. 2001); *Mantech Envtl. Serv., Inc. v. Hudson Envtl. Serv., Inc.*, 152 F.3d 1368, 1373 (Fed. Cir. 1998). A "court should discount any expert testimony that is clearly at odds with the claim construction mandated by the claims themselves, the written description, and the prosecution history, in other words, with the written record of the patent." *Kara Technology Inc. v. Stamps.com Inc.*, 582 F.3d 1341, 1348 (Fed. Cir. 2009) (quoting *Phillips v. AWH Corp.*, 415 F.3d 1303, 1317 (Fed. Cir. 2005)). Nonetheless, when a claim is clear, extrinsic evidence may still be considered to enhance the court's understanding of the technology. *Helmsderfer v. Bobrick Washroom Equipment, Inc.*, 527 F.3d 1379, 1382 (Fed. Cir. 2008). ("A court may look to extrinsic evidence so long as the extrinsic evidence does not contradict the meaning otherwise apparent from the intrinsic record."); *EMI Group N. Am., Inc. v. Intel Corp.*, 157 F.3d 887, 892 (Fed. Cir. 1998).

[CLAIM CONSTRUCTION FOCUSES ON WHAT ONE OF ORDINARY SKILL IN THE ART WOULD HAVE UNDERSTOOD THE CLAIM LANGUAGE TO MEAN.]

The claim construction inquiry is objective and focuses on what one of skill in the art at the time of the invention would have understood the claim language to mean. *Markman v. Westview Instruments, Inc.*, 52 F.3d 967, 986 (Fed. Cir. 1995) (en banc), *aff'd*, 517 U.S. 370 (1996). Accordingly, a technical term used in a patent claim is interpreted as having the meaning a person "experienced in the field of the invention" would understand it to have. *Merck & Co. Inc. v. Teva Pharmaceuticals USA, Inc.*, 347 F.3d 1367, 1370 (Fed. Cir. Oct. 30, 2003); *Hoechst Celanese Corp. v. BP Chem. Ltd.*, 78 F.3d 1575, 1578 (Fed. Cir. 1996). Of course, a patentee is free to act as his or her own lexicographer and may define a claim term differently from its ordinary meaning. *Helmsderfer v. Bobrick Washroom Equipment, Inc.*, 527 F.3d 1379, 1381 (Fed. Cir. 2008). However, if the patentee chooses to act as his or her own lexicographer, the special definition must be "clearly stated within the patent specification or file history." *Id*.

[THE SPECIFICATION IS IMPORTANT WHEN DETERMINING CLAIM INTERPRETATION.]

In fact, the Federal Circuit has determined that "the specification is the single best guide to the meaning of a disputed term and that the specification acts as a dictionary when it expressly defines terms used in the claims or when it defines terms by implication." *Phillips v. AWH Corp.*, 415 F.3d 1303, 1321 (Fed. Cir. 2005) (quoting *Vitronics Corp. v. Conceptronic, Inc.*, 90 F.3d 1576, 1582 (Fed. Cir. 1996). As *Phillips v. AWH Corp.* explained, the specification is not only relevant in claim construction, but it is also usually dispositive. *Id.* at 1315 (where claim construction of the claim term "baffles" was at issue). The Federal Circuit has consistently stressed the importance of the specification in claim construction. *Id.* For instance, the Federal Circuit stated, "[w]hen a patentee explicitly defines a claim term in the patent specification, the patentee's definition controls." *Martek Biosciences Corp. v. Nutrinova, Inc.*, 579 F.3d 1363, 1380 (Fed. Cir. 2009). The Federal Circuit has also held that the patentee's definition governs, "even if it is contrary to the conventional meaning of the term." *Honeywell Int'l Inc. v. Universal Avionics Sys. Corp.*, 493 F.3d 1358, 1361 (Fed. Cir. 2007). Indeed "the best source for understanding a technical term is the specification from which it arose, informed, as needed, by the prosecution history." *Multiform Desiccants*, 133 F.3d at 1478; *see also Metabolite Labs., Inc. v. Lab. Corp. of Am. Holdings*, 370 F.3d 1354, 1360 (Fed. Cir. 2004). Thus, the specification is the primary basis for construing claims. *Phillips*, 90 F.3d at 1315.

[A LIMITATION IN A DEPENDENT CLAIM MEANS THAT THE LIMITATION IS NOT PRESENT IN THE INDEPENDENT CLAIM.]

Under the doctrine of claim differentiation, "the presence of a dependent claim that adds a particular limitation gives rise to a presumption that the limitation in question is not present in the independent claim." *Enzo Biochem, Inc. v. Applera Corp.*, 599 F.3d 1325, 1342 (Fed. Cir. 2010) (quoting *Phillips v. AWH Corp.*, 415 F.3d 1303, 1315 (Fed. Cir. 2005)). Accordingly, the doctrine indicates that where a patent claim does not contain a certain limitation and another claim does, that limitation cannot be read into the former claim in determining either validity or infringement. *SRI Int'l*, 775 F.2d at 1122; *see also D.M.I., Inc. v. Deere & Co.*, 755 F.2d 1570, 1574 (Fed. Cir. 1985).

[CLAIM DIFFERENTIATION CANNOT CONTRADICT THE INTENTION OF THE PATENT DRAFTSMAN.]

The Court of Appeals has held that the doctrine of claim differentiation is not always controlling. In *O.I. Corp. v. Tekmar Co.*,

Inc., 115 F.3d 1576 (Fed. Cir. 1997), the court held "[a]lthough the doctrine of claim differentiation may at times be controlling, construction of claims is not based solely upon the language of other claims; the doctrine cannot alter a definition that is otherwise clear from the claim language, description, and prosecution history." *Id.* at 1582. (citing *Hormone Research Found, Inc. v. Genentech, Inc.*, 904 F.2d 1558, 1567 (Fed. Cir. 1990) (stating that the doctrine of claim differentiation "cannot overshadow the express and contrary intentions of the patent draftsman")). The court stated that the description provided a clear meaning of the inventor's intention and that it "trumps the doctrine of claim differentiation." *Id.*

[EXTRINSIC EVIDENCE SHOULD NOT CONTRADICT THE MEANING OF THE CLAIM.]

Extrinsic evidence should not be used to alter or change the public record which competitors would rely upon to construe the meaning of the claim. This is to enable a competitor to ascertain the scope of a patentee's claimed invention, and thereby design around it. *Vitronics*, 90 F.3d at 1583.

2. Anticipation (35 U.S.C. Section 102)

[AN INVENTION MUST BE NEW OR NOVEL AND NOT ANTICIPATED.]

In general, to be <u>new</u> under Section 102 (also referred to as "novelty" or as not being "anticipated"), an invention may not have been described in a single prior art reference, or the same invention may not have been publicly used or sold more than a year before a patent application is filed. *Glaxo Group Ltd. v. Apotex, Inc.*, 376 F.3d 1339, 1348 (Fed. Cir. 2004); *In re Spada*, 911 F.2d 705, 708 (Fed. Cir. 1990). However, an element may be described in a document which is incorporated by reference in the anticipating reference. *Advanced Display Sys. v. Kent State Univ.*, 212 F.3d 1272, 1282 (Fed. Cir. 2001); *In re Heritage*, 182 F.2d 639, 643 (C.C.P.A. 1950); *Technograph Printed Circuits, Ltd. v. Bendix Aviation Corp.*, 218 F. Supp.1 (D. Md. 1963), *aff'd*, 327 F.2d 497 (4th Cir. 1964), *cert. denied*, 379 U.S. 826 (1964). Moreover, an element not explicitly disclosed in a prior art reference may be disclosed under principles of inherency, particularly where the element not explicitly disclosed would be appreciated by one of ordinary skill in the art and is not a critical aspect of the alleged invention. *Therasense,*

Inc. v. Becton, Dickinson and Co., 593 F.3d 1325, 1332 (Fed. Cir. 2010); *Tyler Refrigeration v. Kysor Indus. Corp.*, 777 F.2d 687, 689-90 (Fed. Cir. 1985). Also, a reference is anticipatory even if a claimed element is not explicitly described if such element was within the knowledge of a person skilled in the art. *In re Robinson*, 169 F.3d 743, 746 (Fed. Cir. 1999); *In re Graves*, 69 F.3d 1147, 1152 (Fed. Cir. 1995).

3. Obviousness (35 U.S.C. Section 103)

[AN INVENTION MUST NOT BE OBVIOUS.]

Under Section 103 of the patent laws, a patentable invention is one which was not obvious to one of ordinary skill in the pertinent scientific field at the time the invention was made. Obviousness is determined using the tests of *Graham v. John Deere Co. of Kansas City*, 383 U.S. 1, 17 (1996):

[DIFFERENCES BETWEEN THE PRIOR ART AND THE CLAIMS SHOULD BE ASCERTAINED.]

Under Section 103, the scope and content of the prior art are to be determined; differences between the prior art and the claims at issue are to be ascertained; and the level of ordinary skill in the pertinent are resolved.

[EVIDENCE OF SECONDARY CONSIDERATIONS SHOULD BE CONSIDERED.]

Objective evidence or the so-called secondary considerations should be considered on the question of obviousness. Such factors as commercial success, long-felt need, new, unexpected and superior properties or advantages, the failure of others, initial expressions of disbelief by experts, industry recognition and copying are to be considered in assessing the validity of a patent. *Stratoflex, Inc. v. Aeroquip Corp.*, 713 F.2d 1530 (Fed. Cir. 1983); *Simmons Fastener Corp. v. Illinois Tool Works, Inc.*, 739 F.2d 1573 (Fed. Cir. 1984); *Gillette Co. v. S.C. Johnson & Son Inc.*, 919 F.2d 720 (Fed. Cir. 1990).

[REFERENCES MAY BE COMBINED TO DETERMINE THAT AN INVENTION IS OBVIOUS.]

Prior art references may be combined to support a finding that an invention is obvious. Something in the prior art as a whole must suggest the desirability, and thus the obviousness, of making the combination. *See Uniroyal*, 837 F.2d at 1051.

Only a reasonable expectation of success, not absolute predictability, is necessary for a conclusion of obviousness. *See In re Merck & Co.*, 800 F.2d 1091, 1097, 231 U.S.P.Q. 375 (Fed. Cir. 1986); *In re O'Farrell*, 853 F.2d 894, 903-904, 7 U.S.P.Q.2d 1673 (Fed. Cir. 1988).

[THE MOTIVATION OR SUGGESTION TO COMBINE SHOULD NOT BE RIGIDLY APPLIED.]

The United States Supreme Court, in *KSR Int'l Co. v. Teleflex Inc.*, 127 S.Ct. 1727 (2007), rejected the rigid application of the Federal Circuit's teaching, suggestion, or motivation to combine known elements (TSM) test. The Court declared that the TSM test, when rigidly applied, was inconsistent with the *Graham* analysis. *Id.* at 1741. The Court determined that in deciding whether the subject matter of a patent claim is obvious, the objective reach of the claim, not the particular motivation or avowed purpose of the patentee, is controlling. *Id.* The question is whether the combination was obvious to a person with ordinary skill in the art. *Id.* at 1742. One of ordinary skill in the art is one of ordinary creativity, and "[w]hen there is a design need or market pressure to solve a problem and there are a finite number of identified predictable solutions, a person of ordinary skill has good reason to pursue the known options within his or her technical grasp." *Id.* As to hindsight bias, the Court cautioned that although there is risk of distortion, "rigid preventative rules that deny factfinders recourse to common sense . . . are neither necessary under our case law nor consistent with it." *Id.* at 1742–43.

VI. ANALYSIS

A. Invalidity Analysis

In our opinion, at least [INSERT CLAIMS] of the [INSERT PATENT] are invalid as obvious in view of the prior art references, as described below.

1. Claim Construction

Prior to assessing the validity of the claims of the [INSERT PATENT], we must first construe those claims to determine their scope.

[INSERT CLAIM] recites:

[INSERT CLAIM] recites:

VII. CONCLUSION OF THE OPINION

For the foregoing reasons, it is our opinion that a court of competent jurisdiction, properly apprised of the relevant facts and law, would hold that [INSERT CLAIMS] of the [INSERT PATENT] are invalid under 35 U.S.C. Section 103(a) and obvious over the above-identified combinations of [INSERT REFERENCES].

Please note that this opinion is based upon our understanding of the facts as set forth herein. If upon your review of this opinion, you become aware of any inaccuracies with respect to our understanding of the facts as set forth herein, or any additional relevant facts, please let us know immediately, so that we can determine whether additional analysis is required.

In considering the opinion rendered herein, you should keep in mind that in a litigation concerning any patent, the outcome is subject to the intangibles presented in all litigations. Such intangibles include, for example, the credibility of witnesses, any additional evidence that might be brought out in litigation, as well as the fact that such litigation would probably involve resolution of a disputed technical fact as to which the parties and their expert disagree.

This letter is a confidential opinion of counsel. It is subject to the attorney-client privilege and the work product immunity, and should not be discoverable in the event of litigation. Nevertheless, the attorney-client privilege and work product immunity can be lost if the confidentiality of the communication is not maintained. For this reason, this opinion should be maintained in a secure location and not be shared with anyone whose duties do not require that he or she be aware of its contents.

If you should have any questions or comments, please do not hesitate to contact me.

Conclusion

Patent lawyers fall into the category of "future" jobs that are needed in the immediate future. The new American Invents Act, which changes how patents are processed and reviewed at the USPTO, has dramatically increased demand for patent lawyers.

In this book, we have addressed the most common issues and pitfalls that come up in patent specifications, amendments, and opinions. The aforementioned documents can be used as templates when drafting your own applications (including claims), responses, and opinions. In addition, you are now equipped to perform a patent search and you now understand the concepts of novelty, non-obviousness, and 101 patentability.

The present guide will certainly give any patent professional an edge over the competition and achieve superior results in the field of patent law.

Index

About the Section

About the ABA Section of Intellectual Property Law

From its strength within the American Bar Association, the ABA Section of Intellectual Property Law (ABA-IPL) advances the development and improvement of intellectual property laws and their fair and just administration. The Section furthers the goals of its members by sharing knowledge and balanced insight on the full spectrum of intellectual property law and practice, including patents, trademarks, copyright, industrial design, literary and artistic works, scientific works, and innovation. Providing a forum for rich perspectives and reasoned commentary, ABA-IPL serves as the ABA voice of intellectual property law within the profession, before policy makers, and with the public.

ABA Section of Intellectual Property Law
Order today! Call 1-800-285-2221
Monday-Friday, 7:30 a.m. – 5:30 p.m., Central Time
or Visit the ABA Web Store: www.ShopABA.org

Qty	Title	Regular Price	ABA-IPL Member Price	Total
_____	ADR Advocacy, Strategies, and Practice in Intellectual Property Cases (5370195)	$139.95	$114.95	$_____
_____	ANDA Litigation (5370199)	$299.00	$249.00	$_____
_____	Careers in IP Law (5370204)	$24.95	$16.95	$_____
_____	Computer Games and Virtual Worlds (5370172)	$69.95	$55.95	$_____
_____	Copyright Remedies (5370208)	$89.95	$74.95	$_____
_____	Distance Learning and Copyright (5370163)	$89.95	$79.95	$_____
_____	Fundamentals of Intellectual Property Law (5370218)	$89.95	$69.95	$_____
_____	Fundamentals of Intellectual Property Valuation (5370143)	$69.95	$49.95	$_____
_____	IP Attorney's Handbook for Insurance Coverage in Intellectual Property Disputes, Second Edition (5370210)	$139.95	$129.95	$_____
_____	IP Protection in China (5370217)	$139.95	$109.95	$_____
_____	A Lawyer's Guide to Section 337 Investigations before the U.S. International Trade Commission, Second Edition (5370203)	$119.95	$89.95	$_____
_____	A Legal Strategist's Guide to Trademark Trial and Appeal Board Practice, Second Edition (5370200)	$159.95	$129.95	$_____
_____	Music & Copyright in America (5370201)	$97.95	$67.95	$_____
_____	New Practitioner's Guide to Intellectual Property (5370198)	$89.95	$69.95	$_____
_____	The Patent Infringement Litigation Handbook (1620416)	$149.95	$129.95	$_____
_____	Patently Persuasive (5370206)	$129.95	$99.95	$_____
_____	Patent Obviousness in the Wake of *KSR International Co. v. Teleflex Inc.* (5370189)	$129.95	$103.95	$_____
_____	Patent Trial Advocacy Casebook, Third Edition (5370124)	$149.95	$119.95	$_____
_____	The Practitioner's Guide to the PCT (5370205)	$139.95	$109.95	$_____
_____	Practitioner's Guide to Trials Before the Patent Trial and Appeal Board (5370209)	$139.95	$114.95	$_____
_____	Pre-ANDA Litigation (5370212)	$275.00	$220.00	$_____
_____	Preliminary Relief in Patent Infringement Disputes (5370194)	$119.95	$94.95	$_____
_____	Right of Publicity (5370215)	$89.95	$74.95	$_____
_____	Settlement of Patent Litigation and Disputes (5370192)	$179.95	$144.95	$_____
_____	Starting an IP Law Practice (5370202)	$54.95	$34.95	$_____
_____	The Tech Contracts Handbook, Second Edition (5370216)	$39.95	$34.95	$_____
_____	Technology Transfer Law Handbook (5370211)	$220.00	$176.00	$_____
_____	Trademark and Deceptive Advertising Surveys (5370197)	$179.95	$134.95	$_____
_____	Trademark Surveys (5370207)	$269.95	$239.95	$_____

* Tax
DC residents add 5.75%
IL residents add 9.25%

Payment
☐ Check enclosed payable to the ABA
☐ VISA ☐ Mastercard ☐ American Express

* Tax $_____
** Shipping/Handling $_____
TOTAL $_____

**Shipping/Handling
Up to $49.99..................... $5.95
$50 to $99.99 .... $7.95
$100 to $199.99................ $9.95
$200 to $499.99.............. $12.95
$500 to $999.99.............. $15.95
$1,000 and above........... $18.95

Name_____

Firm/Organization_____

Address_____

City_____ State_____ Zipcode_____

Phone_____ E-mail_____
(in case of questions about your order)

Please allow 5 to 7 business days for UPS delivery. Need it sooner? Ask about overnight delivery. Call the ABA Service Center at 1-800-285-2221 for more information.

Please mail your order to:
ABA Publication Orders, P.O. Box 10892, Chicago, Illinois 60610-0892
Phone: 1-800-285-2221 or 312-988-5522 • Fax: 312-988-5568
E-mail: orders@abanet.org

Guarantee: If – for any reason – you are not satisfied with your purchase, you may return it within 30 days of receipt for a complete refund of the price of the book(s). No questions asked!

Thank you for your order!

Section of
Intellectual Property Law
AMERICAN BAR ASSOCIATION